Springer Theses

Recognizing Outstanding Ph.I

For further volumes:
http://www.springer.com/series/8790

Aims and Scope

The series "Springer Theses" brings together a selection of the very best Ph.D. theses from around the world and across the physical sciences. Nominated and endorsed by two recognized specialists, each published volume has been selected for its scientific excellence and the high impact of its contents for the pertinent field of research. For greater accessibility to non-specialists, the published versions include an extended introduction, as well as a foreword by the student's supervisor explaining the special relevance of the work for the field. As a whole, the series will provide a valuable resource both for newcomers to the research fields described, and for other scientists seeking detailed background information on special questions. Finally, it provides an accredited documentation of the valuable contributions made by today's younger generation of scientists.

Theses are accepted into the series by invited nomination only and must fulfill all of the following criteria

- They must be written in good English.
- The topic should fall within the confines of Chemistry, Physics, Earth Sciences, Engineering and related interdisciplinary fields such as Materials, Nanoscience, Chemical Engineering, Complex Systems and Biophysics.
- The work reported in the thesis must represent a significant scientific advance.
- If the thesis includes previously published material, permission to reproduce this must be gained from the respective copyright holder.
- They must have been examined and passed during the 12 months prior to nomination.
- Each thesis should include a foreword by the supervisor outlining the significance of its content.
- The theses should have a clearly defined structure including an introduction accessible to scientists not expert in that particular field.

David Blanco-Rodriguez

Modelling and Observation of Exhaust Gas Concentrations for Diesel Engine Control

Doctoral Thesis accepted by
Universitat Politècnica de València, Spain

Springer

Author
Dr.-Ing. David Blanco-Rodriguez
CMT-Motores Térmicos
Universitat Politècnica de València
Camino de Vera s/n
46022 Valencia
Spain
e-mail: dablarod@gmail.com

Supervisor
Prof. Dr.-Ing. Carlos Guardiola
CMT-Motores Térmicos
Universitat Politècnica de València
Camino de Vera s/n
46022 Valencia
Spain

ISSN 2190-5053 ISSN 2190-5061 (electronic)
ISBN 978-3-319-36118-5 ISBN 978-3-319-06737-7 (eBook)
DOI 10.1007/978-3-319-06737-7
Springer Cham Heidelberg New York Dordrecht London

Printed on acid-free paper

Springer is part of Springer Science+Business Media (www.springer.com)

Parts of this thesis have been published in the following journal articles:

Journal Papers

- Galindo J, Serrano JR, Guardiola C, Blanco-Rodriguez D and Cuadrado IG. An on-Engine Method for Dynamic Characterisation of NO_x Concentration Sensors. *Experimental Thermal and Fluid Science*, Vol. 35 n° 3: 470–476, 2011.
- Guardiola C, Pla B, Blanco-Rodriguez D and Cabrera P. A Learning Algorithm Concept for Updating Look-up Tables for Automotive Applications. *Mathematical and Computer Modelling*, Vol. 57 n° 7–8: 1979–1989, 2013.
- Guardiola C, Pla B, Blanco-Rodriguez D, Mazer A and Hayat O. A Bias Correction Method for Fast Fuel-to-Air Ratio Estimation in Diesel Engines. *Proceedings of the Institution of Mechanical Engineers, Part D: Journal of Automobile Engineering*. Vol. 227 n° 8: 1099–1111, 2013.
- Guardiola C, Pla B, Blanco-Rodriguez D and Eriksson L. A Computationally Efficient Kalman Filter based Estimator for Updating Look-up Tables Applied to NO_x Estimation in Diesel Engines. *Control Engineering Practice*. Vol. 21 n° 11: 1455–1468, 2013.
- Guardiola C, Pla B, Blanco-Rodriguez D and Calendini PO. ECU Oriented Models for NO_x Prediction. Part 1: A Mean Value Engine Model for NO_x Prediction. Submitted to *Proceedings of the Institution of Mechanical Engineers, Part D: Journal of Automobile Engineering*.
- Guardiola C, Climent H, Pla B and Blanco-Rodriguez D. ECU Oriented Models for NO_x Prediction. Part 2: Adaptive Estimation by Using a NO_x Sensor. Submitted to *Proceedings of the Institution of Mechanical Engineers, Part D: Journal of Automobile Engineering*.
- Guardiola C, Hoyas S, Pla B and Blanco-Rodriguez D. Solución Analítica de un Filtro de Kalman Estacionario para la Observación de Deriva en Modelos de Emisiones de NO_x en Motores Diesel de Automoción. Submitted to *Revista Iberoamericana de Automática e Informática Industrial*.

Conference Papers

- Desantes JM, Luján JM, Guardiola C and Blanco-Rodriguez D. Development of NO_x Fast Estimate Using NO_x Sensors. *EAEC 2011 Congress*, 2011.
- Guardiola C, Pla B, Blanco-Rodriguez D and Cabrera P. Adaptive Modelling for Automotive Applications. *Modelling for Engineering and Human Behaviour 2011*, 2011.
- Payri F, Guardiola C, Pla B, Blanco-Rodriguez D, Mazer A and Cornette A. Methodology for Design and Calibration of a Drift Compensation Method for Fuel-to-Air Ratio Estimation. *SAE Technical Paper 2012-01-0717*, 2012.
- Payri F, Guardiola C, Pla B, Blanco-Rodriguez D, Mazer A and Cornette A. Fast Estimation of Fuel-to-Air Ratio through the Combination of a Fast Model and an Exhaust Lambda Sensor. *SIA 2012 International Conference*, 2012.

Para mis abuelos Manolo y Feli,
para Mikel, Ama y Aita,
para mi alegría, mi felicidad, Vera
Shine on you crazy diamond

Supervisor's Foreword

It is my pleasure to present Dr. Blanco-Rodriguez's contribution to the *Springer Theses* series. I must admit I feel somehow as a secondary actor in it: I had the honour of advising his work at the CMT-Motores Térmicos institute of the Universitat Politécnica de Valéncia (UPV), Spain, during four long, intense years, and this provided me the opportunity of spending many hours with him discussing the details of the dissertation, defining the approach and going back and forth on the document plot.

Dr. Blanco-Rodriguez has made an extraordinary research work that shows a complete methodology for virtual sensing of engine exhaust variables in turbocharged diesel engines. This is a topic of major relevance for the automotive industry: not only are the engines increasing in complexity, but they exhibit tighter requirements in terms of diagnosis and control. This work comes in time for providing useful solutions for the determination of the exhaust gas composition.

The reader will find a multidisciplinary work that covers the main relevant topics in the field: the use of on-board sensors, the development of models, the design of control-oriented algorithms for virtual sensing and the final successful application on a real diesel engine. The dissertation presents a dedicate chapter reviewing the state of the art while the rest of the chapters focus on the original contributions of this work.

In addition to its clear application-oriented motivation, which will be of high interest for the automotive engineer, the thesis presents relevant developments on observer design and adaptive modelling, which will be of interest to the control engineer, and also for all those facing the always complex activity of confronting different information sources, as sensors and models are.

I anticipate the reader to enjoy it and, hopefully, to extract fresh ideas for its research or for solving real-life problems.

Valencia, March 2014 Carlos Guardiola

Preface

In spite of the economic crisis of transport sectors such as the aerospace or naval, the automotive sector is facing a really challenging time. The trends in this sector during the past years had been focused on minimizing the pollutants, namely NO_x, CO, HC or PM. However, the efforts in the next years are moving towards the CO_2 reduction, and thus minimizing the fuel consumption without worsening the pollutants. In the case of gasoline engines, the emissions problem is relatively solved since years by the three-way catalyst but for diesel, the use of different after-treatment systems as well as new combustion systems and control improvements are necessary. At the same time, the introduction of new fuel types and the development of the electric or hybrid vehicles introduce indeed new possibilities to the engineers.

The diesel engine requires a renewal of the number of information sources available in order to monitor and control all the variables that affect the correct performance of this complex system. A good example is the on-board diagnostics algorithms necessary to monitor the NO_x emissions, required for the urea dosing of the selective catalyst reduction systems or useful for the development of direct NO_x control.

Within this framework, this book intends to give solutions to real problems on the topic of the online estimation of the exhaust engine variables in diesel engines. The reader will find a number of methods and algorithms that cover the main possibilities for inferring the exhaust variables on-board:

On-board sensors
Control oriented models
Data fusion methods for combining the information from sensors and models.

At all times, I have tried not to forget the application and motivation of this work, so that all the methods and solutions presented, have been applied to real engine data for the estimation of NO_x and exhaust lambda. In addition, it is well known that the electronic control unit capacities are limited and thus the computationally efficiency and the simplicity of the methods are mandatory in order to implement the methods on-board. The reader will find then, that these guidelines are kept, so that chapters which cover the engine setup or the applied results are combined with others that directly cover the mathematics of the methods and the development of on-board suitable methods.

From my current position in an automotive company, I am able to test myself the utility of the methods when using them for a big number of tasks in different projects. I would be really satisfied if I could see how this work can be useful for other people in the automotive field or any other field. Furthermore, I am really open to hear the opinion from my colleagues regarding the methods developed here or indeed from some improvement suggestions. I still feel that there are lots of things to do in the field of the adaptive modelling or observation. From my modest contribution, I would like to do my one's bit on this field.

Aachen, May 2014 Dr.-Ing. David Blanco-Rodriguez

Acknowledgments

I have always been, in one way or another, linked to running. When I was younger, I ran in my High-school team (Urdaneta in Loiu). At the same time, I always tried to go faster and faster in all my challenges. In the University, I kept running in everything: I always wanted to be first. However, my great challenge came during the last years, when I decided to move from Sopelana (Basque Country) to Valencia to start my Ph.D. studies. It could be such a coincidence, but it is true, I needed 4 years and 2 months to get to it, as it would be a running Marathon (about 42 kms.)

This has been an endurance race: I enjoyed first kilometres, first years, I got to suffer and had to work quite hard for a long time (suffering what athletics call 'The Wall'), but eventually I kept some strength for sprinting and crossing the finishing line. Now, all the sleepless nights, all the restless weekends and all my endless days in my university office listening to Pink Floyd while working (Another Brick in the Wall?), does not look like being so hard but an experience I've always wanted to live. Whoever that knows me is aware that this was one of my life dreams. I guess that all these stories about great scientists who get amazing achievements, and overall all this need to satisfy my curiosity led me to choose this path.

I would like to give a small brick to these 'pacemakers' who during these years have forwarded me, who have pushing me to get my Ph.D. and who have shown me that with their help, I am able to get everything I wish.

My first greeting is to Dr. Carlos Guardiola, my supervisor and confessor in the Universitat Politècnica de València (UPV). It has been many working hours together, discussing about work and life. You taught me your passion to Research and University, and overall, you showed me the brightest scientific mind I have ever seen. Thanks for all your efforts, and I simply hope that this document and this work should be sufficient to compensate you. I know it is not easy to struggle with my anxiety in going deeper into everything I accomplish; finally I think that I could find the calm (Kalman?) to write a good document.

I would like to express my sincere gratitude to Dr. Lars Eriksson who advised me during nearly 4 months in the University of Linköping in Sweden. I really appreciate all your effort for welcoming me in your country and for spending with me long hours discussing about observers, the Kalman filter and about life in general. I really had a nice time in Linköping.

I would not like to forget my friend and sport teammate Benjamín Pla, who has helped in everything I have needed, both in life and work. Thanks to all my

roommates: Dario, Mike, Pedro and Pozalito. Many thanks to all my working mates, who in one way or another have contributed for me to have nice years in Valencia: Antonio, Raúl, my good friend Andrés 'el Argentino', both Pau's, Alberto (many thanks for your engine figures), Vicent, Roberto, Pedro and a large etc. Thanks Prof. del Re for your attentive review and personal advice.

My dear friends, I apologise for not taking care of you the last years, I promise to be more attentive from now. My special thanks to one of the best, Franck, for being my best friend, I know that everything will go well and Cris and you will be the best parents in the world. Thanks Ivan for not forgetting to call me and for being there, thanks Unai, I really miss you but we'll have good times in the next years. Thanks Natxo, all the 'Armadores de Triana', the people from Aterpe. I know it is hard to understand that I barely have free time when I come back to Bilbao.

Aita, this work is all yours, I know that life puts everyone where they deserve, and you deserve everything. Your patience, social skills and especially your self-confidence are a major learning for me. Whenever I have any problem, I always know that you're there for me. Ama, it is not easy to live far from you, without listening to my first fan. You're the person who shows me that there are more important things than a career, such as love. Mikel, even though you're already 10 cm taller than me, you will always be 'the tiny'. Do not forget that I really love you, that wherever I go I am all the time thinking of you. You are able to overcome any member of our family, if only you put all your effort in your work. You are brilliant and will get everything you want. I will be there to see your success.

This work is sincerely dedicated to my great grandparents, Manolo and Feli, possibly the best persons in the world. I am sure that wherever you are, you are proud of me. Wish you were here.

Thanks Valencia and thanks to you, Vera. You are my pride, my happiness and the girl of my life. It is quite easy to keep on with your smile. You have been my best support during the last long year. I really love you and know you will really succeed. Thanks also to my new family, and especially Quique, MC and Mar to give us the support we needed at this time, I really appreciate you.

And from the views of Barrika, it is time to start a new life with you and looking for our challenges now in Germany; but without forgetting: Run or Die.

Madrid, Sopelana and Aachen
September 2013 and March 2014

Dr.-Ing. David Blanco-Rodriguez

Contents

Part I
Introduction

Chapter 1
Introduction

1.1 Background

The evolution of transportation systems is closely linked with the degree of development of modern societies. As shown in Fig. 1.1, developed countries present passenger car densities around 0.5 cars per inhabitant (Santos et al. 2011; The European Commission 2011), while developing countries, such as China or India around 0.03 (Dargay and Gately 1999; Guzzella and Sciarretta 2007; Payri et al. 2012), although the economic development is expected to make this average grow. The transportation fleet is responsible for about 40 % of the global fuel consumption and the expected growth will rise the number.

1.2 The Emissions Regulation in Diesel Engines

The world attention about environmental protection has resulted in new strict laws which establish the requirements for pollutants emissions, and therefore define priorities in the technology development (Bosch 2011). Traditionally, two types of internal combustion (IC) engines have dominated the transportation market: the spark ignition (SI) or gasoline and the compression ignition (CI) or diesel engines. Particularly in diesel engines, main pollutants are:

- Nitrogen oxides (NO_x) are benefited from high temperatures and a lean combustion. NO_x at the engine exhaust contains NO with a proportion between 70 and 90 % while the rest is mainly NO_2 (Hernández 2004).
- Particulate matter (PM) is a complex aggregate formed by soot, hydrocarbons (HC) resulting from fuel and lubrication and other minor products. The proportion of these components varies with the engine and operating point conditions. PM emissions have an opposite trend with NO_x, and if fuel is burnt appropriately, PM concentration decreases while NO_x increases.

D. Blanco-Rodriguez, *Modelling and Observation of Exhaust Gas Concentrations for Diesel Engine Control*, Springer Theses, DOI: 10.1007/978-3-319-06737-7_1,

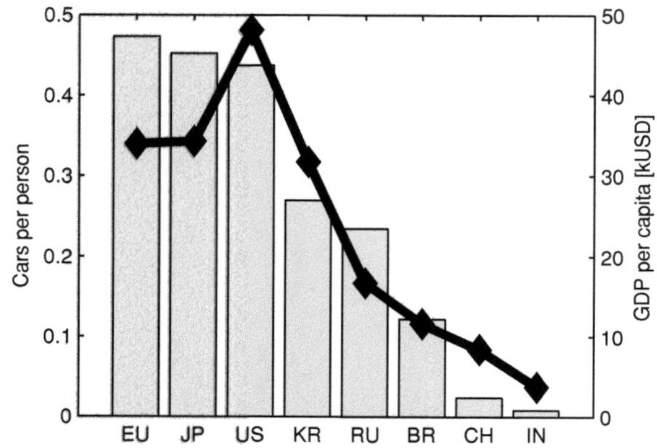

Fig. 1.1 Cars per person (*bars*) and gross domestic product (GDP) per capita (*line*) in some selected countries, adapted from Payri et al. (2012)

- Carbon dioxide (CO_2) formation is proportional to the fuel consumption. Traditionally, CO_2 emissions were not regulated, however they have been driven by the fuel economy demanded by the users. CO_2 is today a main concern for manufacturers not only due to the fuel saving requirements but due to the new regulations on CO_2.[1] The passenger cars are particularly responsible for around the 12 % of the CO_2 emissions caused by humans (Johnson 2010).
- Carbon monoxide (CO) is a sub-product of the combustion (more important in SI than in CI engines) and a dangerous colourless, odourless and tasteless gas. The usual is that CO oxides at the atmosphere for forming CO_2.
- Hydrocarbons (HC) are a product of an incomplete combustion of the injected fuel due to rich conditions and low temperatures that can locally be achieved inside the cylinder. Partially oxidised HC and non-burnt HC are usually included inside this group. The emissions of CO and HC in diesel engines are of less importance than NO_x, PM and CO_2.

Figure 1.2 shows emissions limits for NO_x and PM for the different versions of European legislation standards (The European Parliament and the Council of the European Union 2009) for diesel passenger cars and light-duty (LD) vehicles. Specifically in mobile sources, diesel engines must reduce NO_x emissions by 20 % with regards to EURO 5 and 50 % with regards to EURO 6 if comparing with the previous EURO 4 standard.

Driven by these legislations and the market demands, the diesel engine has substantially changed from the original one designed by Rudolf Diesel. The modern diesel engine comprises of variable geometry turbocharging (Varnier 2012) for increasing the intake air mass flow, common-rail systems for allowing multiple and controlled injections (Schommers et al. 2000), an exhaust gas recirculation (EGR) (Pla 2009) for reducing NO_x and different after-treatment (AT) systems (Johnson 2012) such as the diesel particulate filter (DPF) for trapping soot, the diesel oxidation

---

[1] CO_2 emissions are regulated in Europe since the year 2009 by the European directive 443/2009 (The European Parliament and the Council of the European Union 2009).

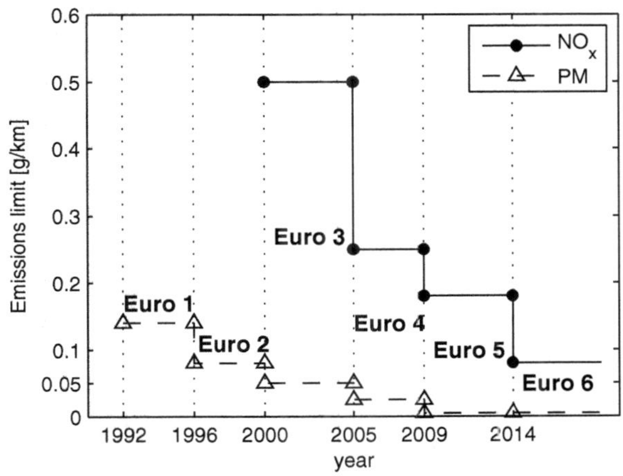

Fig. 1.2 NO_X and PM emission limits for diesel passenger cars and light-duty vehicles (M1) from *Euro 1* to *Euro 6* standards. NO_X limits were first imposed in *Euro 3*. *Euro 6* is to be applied in 2014

catalyst (DOC) for eliminating HC and the selective catalyst reduction (SCR) or the lean NO_X trap (LNT) for removing NO_X. These advances make the diesel engine a sophisticated system that requires an engine control unit (ECU) for an appropriate management and a growing complexity on calibration and implementation.

The SI engine has traditionally solved the emissions problem easier than the CI engine by installing a three way catalyst (TWC), which simultaneously reduces NO_X and oxidises CO and HC. Nevertheless, the fuel efficiency is lower in SI than in CI engines. In the last years, and with the idea of taking advantage of the best properties of CI and SI engines, manufacturers and researchers have put a big effort in developing a more efficient and environmentally sustainable engine. These concepts usually lie on low temperature combustions (LTC) (Lu et al. 2011). Alternatively, hybrid-electric and pure electric power sources are also becoming an option for the present and future vehicle.

Whatever the engine selected (IC, electric or hybrid), the appropriate management of all subsystems merits a special focus on the ways of retrieving information about the states (i.e. variables that determine the behaviour of the engine). The work presented in this dissertation is focused on the estimation of the exhaust gas concentrations of diesel engines, whose motivation is further presented in the next section.

1.3 The Need of Information in Diesel Engines

The advancements in the diesel engines must be in the direction of guarantying low emissions while achieving high efficiency, performance and reliability. These purposes require not only efforts on the engine design and performance, but also in the development of reliable measurement systems to get information about the processes, which is needed for the implementation of control strategies, paying a special attention to the transient operation. In such sense, standards also specify on-board diagnostics procedures (OBD) that, based on models and measured data, are capable of detecting faults on engine or vehicle system. Driver must be notified when emissions are out of the limits with the malfunction indicator lamps (MIL) and the ECU must correct this fault or recommend what to do for a correct and safe actuation.

The diesel engine itself may be split in two main systems: the fuel and the air paths. The fuel path is controlled by the measured rail pressure and the estimated injection fuel rate. Globally, the engine is controlled for delivering a given referenced torque on the wheels. However, no variable is measured downstream the combustion chamber and that means that engine management is not based on final measurements, and emissions limits or performance are tracked by feed-forward controllers with closed loop control on intermediate variables.

Regarding the emissions control in LD diesel engines without SCR and LNT, the common layout is installing no physical nor virtual sensor for tracking pipe-out emissions; the engine is calibrated offline before going to market. During the calibration procedure, the emissions are measured by laboratory sensors with limited dynamic capabilities. Furthermore, standard procedures are often based on steady-state measurements and indeed, dynamic homologation cycles, such as the new European driving cycle (NEDC) (see Chap. 3), are not representative of real-life driving cycles (Rubino et al. 2008). Finally, hardware unit-to-unit dispersion (engine, sensors, etc.), system ageing and the effects of external variables (temperature, pressure, humidity, etc.) make this procedure does not ensure that the engine keeps working as when it was calibrated.

Narrow band lambda sensors for measuring lean or rich mixture have been installed in SI engines from 1968 (for controlling the TWC) and wide band lambda sensor for providing full resolution oxygen concentration can be found in some diesel engines from the end of the 90s. Anyway, the exhaust oxygen measurement in diesel engines is not standard and is not usually utilised for controlling the air path (except for detecting injector drifts).

Furthermore, it is less usual finding on-board sensors for measuring engine-out emissions in LD diesel engines. Different reasons can be stated:

- Exhaust gas concentration sensors (mainly NO_x and soot) with a sufficient small size and acceptable accuracy for on-board measurements have not been available until the last decade. More recently, on-engine NO_x can be measured with NO_x sensors based on the ZrO_2 planar technology (Kato et al. 1996; Zhuiykov and Miura 2007). On-board soot sensors are also starting to be marketed.
- The high cost of these on-board sensors with respect to the traditional ones used in automotive until EURO 5 standards [the cost can be several times higher according to Fleming problem of the online estimation (2001)].
- Even though the steady-state accuracy of these sensors is acceptable, the dynamic performance is still problematic due to the slow and filtered responses in the order of 1 s, which prevent from using them with real-time purposes.

Anyway, these sensors are being gradually introduced in the market, enhanced by the development of after-treatment devices.[2] Admittedly, some luxury cars have already installed NO_x sensors some years before. However, the problems derived of using the raw output signal in real-time functions must be addressed.

Models may be used as an alternative to sensors, since they can reproduce fast transients and present offline prediction capabilities (Wahlström and Eriksson 2011; Schilling and Amstutz 2006). Nevertheless, models suffer from drift depending on time and engine operating conditions, and uncertainties linked with the model structure or non-considered inputs. In commercial engine ECUs, the prevailing approach is to use look-up tables to model nonlinear and operating point dependencies because of the simple programming and easy comprehension. An intensive test campaign should be made for filling the tables with a big number of steady-states for offline calibration and a limited number of dynamic tests for validation. Anyhow, the drift problem cannot be eliminated unless sensors with acceptable accuracy are used.

Two main options can be considered for cancelling drift: modelling the bias as an additive or multiplicative factor to the model, or recalibrating the model. For both adaptive options, observers can be designed. For the former, the bias can be tracked by designing a proper drift correction observer, while for the latter, the model parameters can be updated for minimising the errors.

For all those techniques and cases, the Kalman filter (KF) (Kalman 1960) is a standard tool for observing variables. The multiple possibilities for designing observers make feasible to estimate drift, correct variables and indeed calibrate and update tables and models, all in an online basis required for diesel diagnostics or control. Anyway, computational limitations should be issued, and optimized algorithms and simplifications are crucial, jointly with the robustness of the algorithms.

1.4 Scope of the Work

The dissertation covers the problem of the online estimation of diesel engine exhaust concentrations of NO_x and the fuel-to-air ratio or richness (λ^{-1}). Two information sources are utilised:

- On-board sensors for measuring NO_x and λ^{-1}, and
- Control oriented models (COM) in order to predict NO_x and λ^{-1}.

The evaluation of the static accuracy of these sensors is made by comparing the outputs with a gas analyser, while the dynamics are identified on-board by performing step-like transitions on NO_x and λ^{-1} after modifying ECU actuation variables. The former are produced by provoking sharp changes in the start of injection (SOI) while the latter by sharp changes in the injected fuel. With a sufficient data set, the response time and delay of the sensor are identified.

[2] In HD engines it is usual finding NO_x sensors for a correct urea dosing in SCR systems.

Afterwards, control oriented models for estimating NO_x and λ^{-1} are proposed. Regarding the λ^{-1} prediction, the computation is based on the relative fuel-to-air ratio, where fuel comes from an ECU model and air mass flow is measured by a sensor. For the case of NO_x, the model is set-point relative and based on look-up tables fitted for representing nominal engine emissions with an exponential correction based on the intake oxygen variation. Different correction factors for modelling other effects such as the thermal loading of the engine are proposed. The model is able to predict NO_x with a low error and a simple structure.

Regardless of using models or sensors, model drift and sensor dynamic deficiencies affect the final estimation. In order to solve this problem, data fusion strategies are proposed by combining the steady-state accuracy of the sensor and the fast estimation of the models. In a first approach, a drift correction model tracks the bias between the model and the sensor but keeping the fast response of the model. A KF is used for observing the bias, acting as a high-pass filter for the model and a low-pass filter for the sensor. In a second approach, the updating of look-up tables by using observers is coped with different versions based on the extended Kalman filter (EKF). Particularly, a simplified KF (SKF) allows to observe the parameters with a low computational effort.

Finally, the methods and algorithms developed in this work are combined and applied to the estimation of NO_x and λ^{-1} by drift correction algorithms and by developing adaptive look-up tables. The methods developed in this work are applied to NO_x and λ^{-1}, however these could be employed for other variables such as soot, or engine parameters such as the volumetric efficiency.

1.5 Objectives

The present work provides solutions and ideas in the field of adaptive estimation and drift correction of signals and models. Particularly, the following are the main objectives of the present work:

- The introduction of a novel method for the online characterisation of NO_x sensors (Chap. 3).
- The development of a simple ECU-oriented model for diesel NO_x prediction (Chap. 4).
- The design of a simplified version of the EKF with similar accuracy as the standard KF, but that requires a much lower memory resources and calculation time, for updating look-up tables and models (Chap. 5).
- The development of a full methodology covering design, tuning and on-engine implementation of observers for the estimation of NO_x and λ^{-1} (Chaps. 5 and 6).
- The design of a set of observers for updating an ECU-oriented model based on physical and virtual sensors and applied to the NO_x prediction (Chap. 6).

1.5.1 Methodology

This document is split up among different parts, which at the same time can contain more than one chapter. Part I covers the first two chapters and presents the introduction and the state of the art in estimating exhaust gas concentrations. Part II comprises Chaps. 3 and 4, and presents the methodology for on-board measuring and modelling of gas concentrations. Part III, which covers Chaps. 5 and 6, introduces filtering algorithms for fusing the data coming from sensors and models. Furthermore, it remarks the design of computationally efficient methods for its on-board implementation. Part IV comprises Chap. 7 and presents the conclusions and future works. Finally, the document also includes appendices and the curriculum vitae of the author. In addition, every chapter has its own bibliography for the sake of clarity.

In the following, the work methodology is sketched up and split up among the aforementioned chapters.

Chapter 1 has introduced the background and motivation of the present work, centred on the need of information and sources for obtaining in CI engines in order to estimate exhaust engine variables such as NO_x and λ^{-1}. In addition to this, the objectives and methodology of the work have been presented.

Chapter 2 emphasises the need of information for a proper control of the diesel engine and reviews sensors and models for estimating gas concentrations in diesel engines, while also presents different methods for fast estimation of variables. These methods are focused on the drift correction and updating algorithms for look-up tables.

Chapter 3 describes the system set-up including the engine characteristics, which in the present work has been a 2.2 L turbocharged DICI engine, and the different engine tests performed on that engine. Furthermore, a novel procedure for the online characterisation of NO_x sensors based on SOI variations is depicted. Characterisation of λ^{-1} output from NO_x sensors is also presented. This is a key point for the application of the algorithms developed in subsequent chapters, as the knowledge on the sensor response greatly affects the results.

Chapter 4 presents the design and validation results of ECU-oriented models for λ^{-1} and NO_x prediction. The first is just based on the calculation of the fuel-to-air ratio by the injected fuel mass flow and the air mass flow signals from the ECU while the second is based on a nominal set-point relative fitting of the NO_x with a series of corrections for accounting with variations on λ^{-1}, temperatures and other signals. The NO_x model combines look-up tables with physical-based equations and is designed for being implemented on commercial ECUs.

Chapter 5 presents two data fusion strategies for the fast estimation of variables when a steady-state accurate sensor and a fast model are presented. The first is based on drift correction algorithms in order to track the bias between a model and a sensor. In order to observe the bias, a steady-state KF is designed and the observer tuning is also discussed.

The second strategy comes from that look-up tables are often used in automotive control systems and there is a need for systematic methods that can estimate or update them on-line. According to that, learning algorithms for the online updating of look-up tables are proposed. Based on the EKF for observing look-up tables parameters (Höckerdal et al. 2011), two different versions that keep some of the properties of the original KF but optimise the computation are designed.

The methods presented in the Chap. 5 can be applied to different engine variables. In this work, Chap. 6 applies them to the fast estimation of λ^{-1} and NO_x. In order to do that, different algorithms based on the dynamic estimation methods of Chap. 5 are proposed.

Afterwards, a full methodology for the fast estimation of λ^{-1} is presented. This part covers the methodology, the implementation of robustness conditions for coping with signal uncertainties and proposes an adaptive look-up table for modelling operating point dependency of the bias. Finally, experimental results are presented.

The second part of the chapter is centred on the fast estimation of NO_x by designing different algorithms. First, the evaluation of the methods for updating look-up tables designed in the Chap. 5 is validated by using real engine data for the NO_x estimation. Furthermore, single 2D adaptive look-up tables scheduled by injection and speed are proposed for modelling slow varying cycles. Second, different set of observers are proposed in order to update parameters on the NO_x model developed in the Chap. 4 and for minimising the errors in the model inputs when sensor signals are available. To conclude, the estimation of the actual NO_x, solving the problems of filtering and delay of sensors, is proposed.

Finally, Chap. 7 presents the conclusions and the future works derived from this dissertation.

References

Bosch R (2011) Automotive handbook. Bosch handbooks, 8th edn. Robert Bosch GmbH, Stuttgart

Dargay J, Gately D (1999) Income's effect on car and vehicle ownership, worldwide: 1960–2015. Transport Res Part A Policy Pract 33(2):101–138

Fleming WJ (2001) Overview of automotive sensors. IEEE Sens J 1(4):296–308

Guzzella L, Sciarretta A (2007) Vehicle propulsion systems. Introduction to modeling and optimization, 2nd edn. Springer, Berlin. ISBN 978-3-540-74691-1

Hernández L (2004) Desarrollo de una Metodología para la Predicción y Optimización de Emisiones Contaminantes y Consumo en Motores Diesel de Automoción Mediante Redes Neuronales Artificiales. PhD thesis, Universitat Politècnica de València. Departamento de Máquinas y Motores Térmicos

Höckerdal E, Frisk E, Eriksson L (2011) EKF-based adaptation of look-up tables with an air mass-flow sensor application. Control Eng Pract 19:442–453

Johnson TV (2010) Review of CO_2 emissions and technologies in the road transportation sector. SAE technical paper 2010–01-1276

Johnson TV (2012) Vehicular emissions in review. SAE technical paper 2012–01-0368, vol 5, no. 2

Kalman RE (1960) A new approach to linear filtering and prediction problems. J Basic Eng 82 (35–45)

Kato N, Nakagaki K, Ina N (1996) Thick Film ZrO_2 NO_x Sensor. SAE technical paper 960334

Lu X, Han D, Huang Z (2011) Fuel design and management for the control of advanced compression-ignition combustion modes. Prog Energy Combust Sci 37:741–783
Payri F, Luján JM, Guardiola C, Pla B (2012) A challenging future for the IC engine: new technologies and the control role. In: Keynote in ECOSM 2012 Workshop on engine and powertrain control, simulation and modeling
Pla B (2009) Análisis del Proceso de la Recirculación de los Gases de Escape de Baja Presión en Motores Diesel Sobrealimentados. PhD thesis, Universitat Politècnica de València. Departamento de Máquinas y Motores Térmicos
Rubino L, Bonnel P, Hummel R, Krasenbrink A, Manfredi U, De Santi G (2008) On-road emissions and fuel economy of light duty vehicles using PEMS: chase-testing experimente. SAE technical paper 2008–01-1824
Santos A, McGuckin N, Nakamoto HY, Gray D, Liss S (2011) Summary of travel trends: 2009 national household travel survey. US Department of Transportation, Federal Highway Administration
Schilling A, Amstutz A, Onder CH, Guzzella L (2006) A real-time model for the prediction of the NO_x emissions in DI diesel engines. In: Proceedings of the 2006 IEEE international conference on control applications, Munich, Germany
Schommers J, Duvinage F, Stotz M, Peters A, Ellwanger S, Koyanagi K, Gildein H (2000) Potential of common rail injection system for passenger car DI diesel engines. SAE technical paper 2000–01-0944
The European Commission (2011) Energy. Publications Office of the European Union, Transport and Environment Indicators
The European Parliament and the Council of the European Union (2009) Regulation (EC) No. 443/2009 of the European Parliament and of the Council of 23 April. Official Journal of the European Union
Varnier O (2012) Trends and limits of two-stage boosting systems for automotive diesel engines. PhD thesis, Universitat Politècnica de València. Departamento de Máquinas y Motores Térmicos
Wahlström J, Eriksson L (2011) Modelling diesel engines with a variable-geometry turbocharger and exhaust gas recirculation by optimization of model parameters for capturing non-linear system dynamics. Proc Inst Mech Eng [D]: J Automobile Eng 225(7):960–986
Zhuiykov S, Miura N (2007) Development of zirconia-based potentiometric NO_x sensors for automotive and energy industries in the early 21st century: what are the prospects for sensors? Sens Actuators B: Chem 121(2):639–651

Chapter 2
Exhaust Gas Concentrations Estimation in Diesel Engines

2.1 Introduction

Driven by the advancements in diesel engines and especially due to the stringent regulations in the recent years that have enhanced the introduction of the after-treatment (AT) systems, the control and diagnostic logics in the modern ECU are accomplishing a renewal. The typical configuration of a short route EGR and a variable geometry turbine (VGT) shall be extended to include dual EGR loops, and different systems for eliminating particles, hydrocarbons and NO_x in the exhaust. This modernisation must be run jointly with the improvements in the ways of retrieving information from the states of the engine, particularly from engine-out concentrations. This chapter gives an overview on the actual state on technologies related with diesel engine and the different sources of information from the exhaust gas concentration on diesel engines: sensors, models and adaptive estimators. A specific discussion about the λ and NO_x estimation in CI engines is made due to their relevance in the present dissertation.

2.2 Diesel Engine Subsystems

There exists different possible diesel engine layouts, but they all share at some point the following subsystems (Bosch 2011): the fuel path, the air path, the after-treatment and the control system.

2.2.1 *The Fuel Path System*

The common-rail (CR) system and the direct injection (DI) is standard in current diesel engines (Schommers et al. 2000). The main advantages of the CR against

D. Blanco-Rodriguez, *Modelling and Observation of Exhaust Gas Concentrations for Diesel Engine Control*, Springer Theses, DOI: 10.1007/978-3-319-06737-7_2,

other systems, such as the distributor pump and the unit pump injector, are the multiple injections and the control flexibility. A pressurised deposit monitored by a rail pressure sensor is capable of maintaining the pressure highly constant during the injection.

In the ECU, the total injected fuel mass (m_f) is modelled by a look-up table function of the injection duration (t_{id}) and the rail pressure (p_{rail}). The SOI actuation (u_{soi}) is determined by a calibrated look-up table as function of speed (n) and the desired m_f and is measured in degrees with respect to the top dead centre (TDC). In the case of sharp load steps and in order to avoid the excessive emissions of HC and soot, the smoke limiter function limits m_f until the air path responds and supplies the air mass flow ($\dot{m}_a$) required for the combustion, as depicted later in Fig. 2.2.

$$m_f = f(t_{id}, p_{rail}) \tag{2.1a}$$

$$u_{soi} = f(n, m_f). \tag{2.1b}$$

2.2.2 The Air Path System

In a commercial CI engine, the following subsystems can usually be found in the air path: the air filter (in order to clean the entering air, although causing a minimum pressure drop in the line), a (single or double stage) turbocharger (TC) including an intercooler for cooling the compressed air (thus increasing the density), the intake and exhaust manifolds, the EGR loops (high and low pressure loops) and the pipe-out line (the after-treatment devices are considered as an independent subsystem because of the different possibilities existing today, even though they are really connected to the pipe-out line).

A single TC is the common layout in commercial engines, but some mount also dual stage TC (Varnier 2012). The main function is increasing the intake air mass flow by increasing the density with a compressor usually powered by a turbine shaft located at the engine exhaust (Saidur et al. 2012). One of the main drawbacks of the system is the turbolag of the TC which is associated with the inertial response of the turbo shaft to accelerations, especially at low speed and low loads, when the exhaust energy is quite low (Park et al. 2010). In order to get the maximum efficiency of the compressor and turbine, two types of control are usual: waste-gate (WG) valves and a variable geometry turbine (VGT), also known as variable nozzle turbine (VNT). The WG configuration is based on the use of a fixed geometry in the turbine with a valve that acts bypassing the flow minimising the effective exhaust flow through the turbine. This solution has practically been replaced by the use of the VGT, which is based on movable nozzles that vary the turbine work depending on the operating point. The VGT allows to increment the engine performance reducing the specific fuel consumption and also adds a major flexibility to the control, admittedly that increments the related complexity and calibration effort. In a dual stage TC, there

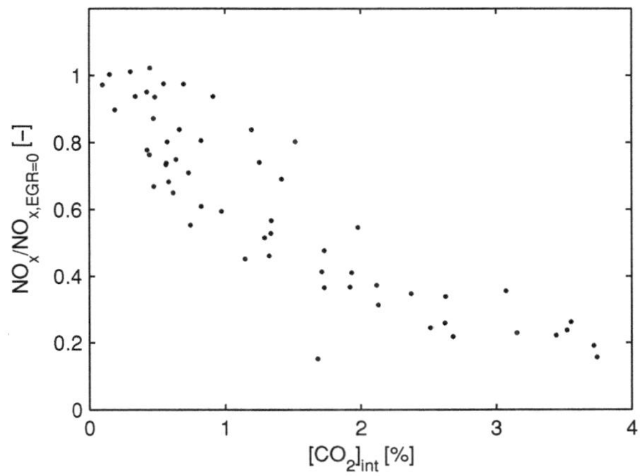

Fig. 2.1 Nominal NO_x emissions depending on $[CO2]_{int}$ for nominal tests in the CI engine explained in Chap. 3. Values are normalised with the maximum NO_x value for each operating point, which corresponds to $u_{egr} = 100$ (fully closed) for each pair of $[n, m_f]$

exists a wide variety of possibilities of combining WG, VGT and fixed geometries in the high pressure (HP) and low pressure (LP) turbines (Galindo et al. 2009).

The exhaust gas recirculation (EGR) is the most widely extended NO_x reduction system (Pla 2009) and is used in CI engines since the middle of the 90s. The basic idea is recirculating a portion of the exhaust gas to the intake. The principal effect of the EGR system is reducing the effective O_2 (increasing effective $[CO2]_{int}$) at the intake and then diminishing the peak temperature in the cylinder and thus reducing NO_x as it is shown in Fig. 2.1.

The EGR rate in steady-state operation can be defined as the quotient between the total EGR mass flow $\dot{m}_{egr}$ and the intake mass flow ($\dot{m}_{int}$)

$$\dot{m}_{int} = \dot{m}_a + \dot{m}_{egr} \tag{2.2a}$$

$$EGR = \frac{\dot{m}_{egr}}{\dot{m}_{int}} \tag{2.2b}$$

The work by Ladommatos et al. (1996a, b, 1997a, b, 2000) is a good reference for understanding the effects of the EGR in the combustion and emissions. In addition to the internal EGR (inert gas fraction that stays in the cylinder after the combustion), the external EGR can comprise high pressure EGR (HPEGR) and low pressure EGR (LPEGR) loops. The HPEGR is based on the extraction of a portion of the exhaust gas upstream of the turbine and driven normally to the intercooler output (hot EGR). The LPEGR extracts a portion of the exhaust gas at some point downstream of the turbine (it is usual to locate it downstream of the diesel particulate filter for avoiding damage in the compressor) and is guided at the compressor inlet (after the air filter). Besides the installation of valves for controlling the flow, coolers are installed for increasing the flow density.

The HPEGR, namely EGR for simplification, is the standard in CI engines due to the simple layout and control: the pressure differences between exhaust and intake facilitate that gas goes to the intake with some exceptions, which can occur for high EGR rates, problem that can be solved e.g. by introducing a throttle valve (van Nieuwstadt 2003). Even though the LPEGR did not enter into massive production, there exists today a certain interest in combining both loops for further reduction of emissions that can push up to relax the AT efficiencies (Millo et al. 2012; Desantes et al. 2013) and thus the cost of complex systems.

2.2.2.1 EGR/VGT Control

The air path is managed independently of the fuel injection system. The effects of the injection settings are considered in the air path as instantaneous inputs for a multiple input multiple output (MIMO) control problem. More concretely, the fuel injection quantity (m_f) and the engine speed (n) are used as scheduling variables for defining the set-points for the EGR and VGT actuation. Then, for a common layout of HPEGR and VGT (or WG), the inputs are the EGR valve position (u_{egr}) and the turbine actuation (u_{vgt} or u_{wg}), besides other actuators such as valves for bypassing the flow over the EGR cooler ($u_{bp,egr}$) if necessary. If the TC is dual stage, it is usual that only one of the turbines can be actuated. For other layouts, different control strategies can be designed, e.g. with a twin turbo sequential parallel configuration, an extra variable is necessary in order to coordinate the switching (Galindo et al. 2009). Varnier (2012) gives a complete review on boosting technologies and its associated control.

The EGR/VGT control strategy basically consists on two maps for determining the set-points for the air mass flow ($\dot{m}_a$) and the boost pressure (p_{boost}), while u_{egr} and u_{vgt} are commanded for reaching these set-points (Guzzella and Amstutz 1998). The coupling of the EGR/VGT system is avoided by defining control regions, with and without EGR. When engine is working in the EGR area, u_{egr} is controlled in closed loop for reaching the required air mass flow set-point scheduled by m_f and n. The VGT is commanded in open loop by a feed-forward controller that imposes a given u_{vgt} depending also on m_f and n. This structure avoids the problem of the EGR/VGT coupling that can be exemplified as follows: when the EGR valve is opened, a closing on turbine nozzles increments the exhaust backpressure and the turbine regime, this effect produces a higher recirculation of EGR reducing the effective air mass flow at the intake (the EGR mass flow replaces the fresh air). However, when the EGR valve is fully closed, a closing of the turbine nozzles produces an increase of the intake air mass flow. This occurs for medium-high engine loads and u_{vgt} is controlled in closed loop, while the EGR valve keeps closed. The control objective is then reaching the intake air mass flow set-point. Figure 2.2 schematises the air path and the smoke limiter controller.

Furthermore, some corrections or maps switching are applied depending on combustion modes or for cold starting strategies (the references for injection and air path are modified). The engine coolant temperature (T_{cool}) is also used for corrections and for defining the cold starting strategies.

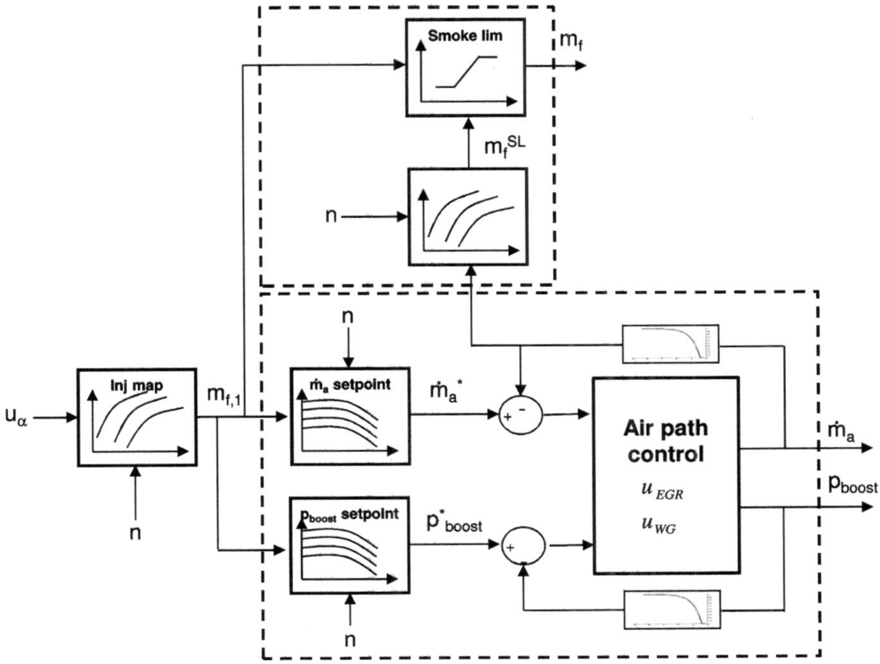

Fig. 2.2 Smoke limiter and air path control blocks

An alternative to the commercial air path control is using virtual and physical sensors or including adaptive strategies by feeding back the controllers with direct measurements of exhaust conditions, e.g. exhaust gas concentrations whose estimation is further discussed in Sect. 2.3. Then, model predictive control (MPC) (Tschanz et al. 2012; Ortner and del Re 2007) or other strategies (Stefanopoulou et al. 2000) can be exploited for such control.

2.2.2.2 Dual EGR/VGT Control

When using a dual EGR loop, in addition to the control variables presented for the single loop, the following inputs are necessary: LPEGR valve (u_{lp}), intake throttle valve (u_{it}) and/or the backpressure valve positions (u_{bp}). The total EGR rate is calculated as in (2.2b) but considering mass flows of the low and high pressure loops respectively $\dot{m}_{lp}$ and $\dot{m}_{hp}$

$$\dot{m}_{egr} = \dot{m}_{lp} + \dot{m}_{hp} \tag{2.3}$$

Depending on the engine conditions and control needs, the EGR split ($\dot{m}_{lp}/\dot{m}_{hp}$) is selected, i.e. the set-points for $\dot{m}_{lp}$ and $\dot{m}_{hp}$. The control is similar to that of the single EGR/VGT control but designing an appropriate algorithm for defining the

Table 2.1 Approximate characteristic response times for the main actuation variables in a CI engine

Command	Response time
u_{prail}, u_{soi}, u_{id}	100–600 μs
u_{hp} (u_{egr})	0.2–0.5 s
u_{lp}	0.2–1 s
u_{vgt}, u_{wg}	0.5–1.5 s

EGR split (Shutty et al. 2007). A common solution in steady-state is controlling u_{lp} and if necessary opening u_{hp} for reaching the $\dot{m}_a$ set-point (EGR is not directly measured). Indeed, the LP loop itself might be sufficient for getting the required $\dot{m}_a$ set-point. For transient operation, the discussion must evaluate the characteristic response times of the systems: the HPEGR line is faster than the LPEGR line due to the higher pressure differences and the shorter length. This produces that different trade-offs between u_{hp} and u_{lp} may be stated during transients. An intrinsic benefit of using a dual loop EGR is that for the EGR engine area, the total $\dot{m}_{egr}$ quantity may be increased leading to further reductions on NO_x, which can alleviate the needs of AT or could move the deNO_x[1] trade-off to select a lean NO_x trap (LNT) rather than a selective catalyst reduction (SCR) system. Model based and optimal approaches to the control problem of dual EGR loops can be found in (Haber and Wang 2010; Chauvin et al. 2011; Desantes et al. 2013).

2.2.2.3 Control of the Air and Fuel Paths

Some authors have proposed a joint control of the air and fuel paths for optimising the engine transient operation. For that, fast acting variables linked to the fuel path (u_{soi} or p_{rail}) are able to track NO_x or soot while u_{egr} and u_{vgt} are arranged to improve the engine transient response while keeping the engine operation (ensuring the torque on wheels) without exceeding emission limits. This can be made by feeding back the controllers with real time measurements of exhaust gas concentrations (typically NO_x and soot, or exhaust oxygen) (Deng et al. 2012; Tschanz et al. 2012). Table 2.1 sums up the characteristic response times for step variations in the inputs, showing how the associated fuel path responses are quite faster than the air path ones. Furthermore, the response times of the TC are in general higher than those related to the EGR loop due to the turbolag problem.

Some authors have proposed a transient operation optimization of the engine management strategy (EMS), see e.g. (Alberer and del Re 2009b). However, manufacturers are still reluctant to change the actual implementation in the fuel and air path system: the EMS is confided to fixed calibrations and optimised for steady-state operation with a validation in transient operation without feedback on exhaust engine states.

[1] The term deNO_x refers in general to an AT system for reducing NO_x, see Sect. 2.2.3.

2.2.3 The After-Treatment Systems

The new stringent regulations have forced to get an impressive development in diesel engines, and particularly with the after-treatment devices (AT); Johnson (2012) and Twigg (2007) present complete reviews on emissions control and available AT technologies. The standard configuration for AT in diesel engines for EURO 6 will comprise

- a diesel oxidation catalyst (DOC) for three main functions: oxidising CO and HC, heating the exhaust gas and oxidising NO to NO_2, which is useful in the DPF regeneration and also facilitates the SCR performance;
- a diesel particulate filter (DPF) for eliminating soot; and
- a deNO$_x$ system for reducing NO_x that it is usually a selective catalyst reduction (SCR) or a lean NO_x trap (LNT).

The DPF and LNT action is based on substrate materials that store the pollutants through mechanisms or chemical principles. When the system is full, a regeneration by means of post-injections and thus increasing exhaust heat is necessary. The DOC system is built from ceramic and catalytic noble metals, such as platinum or rhodium, for enhancing the oxidation of CO and HC as well as NO to NO_2. SCR systems eliminate NO_x by dosing urea and water (principle component is ammonia) and reacting with NO_x for obtaining N_2 and water, where a dedicated control management is necessary for optimising its dosing (Chi and Da Costa 2005; Roberts 2011). The SCR requires an specific installation for the urea storage that increases the cost and weight, and for this reason, the SCR is more extended in HD than in LD. Major concerns in SCR are the light-off related with the minimum temperature for guarantying the efficiency of the system (critical when the engine is cold or it is working in low-load operation); and the ammonia slip, due to the excessive injection of urea, which does not react with NO_x, constituting an economical cost and affecting the output of the NO_x sensors based on ZrO_2 (see Sect. 2.3.1).

DPF and DOC are standard in EURO 5 cars, and deNO$_x$ systems will also be with the application of EURO 6. Anyway, the trade-off between SCR and LNT is not solved (Bickerstaffe 2009), and different factors influence the final decision. Thinking of future regulations, manufacturers struggle between getting lower removal efficiencies (around 85 %) by using LNT or allowing the engines to work on much higher efficiencies and then higher NO_x production but with the use of a SCR with a removal efficiency nearly 98 % (Johnson 2012).

The selection of the layout of the ATs in the engine is another important trade-off and influence the correct operation of the systems. Some possibilities are later shown in Fig. 5.1; however this is a trending discussion. It seems logical that the DOC should be located first because it is required for a proper performance of DPF and SCR, but the trade-off between DPF and SCR or LNT present advantages and drawbacks. Indeed, there are some research on moving the DPF upstream of the turbine for preventing turbine and compressor damage (if it is located before the EGR loop) (Payri et al. 2011b) or doing something similar with the DOC

(Carberry et al. 2005). There are also some alternatives that combine some of the systems for reducing the packaging and final weight (Frobert et al. 2012).

With respect to the feedback, the knowledge of the concentrations upstream and downstream of the AT systems allows to diagnose the performance, and is required for direct control of the final emissions. The temperature also influences the regeneration of the DPF and LNT as well as it is critical for the SCR light-off (key reactions are inhibited at low temperatures). The pressure drop in the line could also be used for detecting failures or when the DPF or LNT are full. The decision on the sensor set for a proper control and diagnosing is a challenging problem that requires a specific study out of the scope of this work. The use of sensors and models for estimating exhaust gas concentrations, especially NO_x and exhaust oxygen, is discussed in Sect. 2.3.

2.2.4 The Control System

As presented in previous sections, the modern diesel engine has evolved until reaching today an important and increasing complexity, which is reflected with the introduction of different subsystems for improving the performance and minimising the emissions. The vehicle electronics in the 80s were reduced to the radio plus simple engine controllers. At the same time, these were based on mechanical and hydraulic devices with a simple electronic assistant, and dedicated to specific functions such as the ignition in the case of SI engines. However, the evolution of the engines has been closely related with the rise of complexity of the modern embedded electronics in automotive, as it is well depicted in (Ebert and Jones 2009). Nowadays, safety performance, fuel efficiency and comfort are major demands and these are achieved by installing proper control units. Modern cars have between 20 and 70 control units with a size of 10^8 in object instructions, while in the first 90s this was in the order of 10^6, according to Ebert and Jones (2009).

From all the control units in the vehicle, the ECU is in charge of managing the engine. Others work with the braking systems, the air conditioning in the cabin or the gear shifting management (Bosch 2011). In the beginning of the 00s, the ECU EDC16 model by Bosch for diesel engines changed the view of the diesel control by using the final torque demand on the wheels as the feedback for a proper air and fuel path management. This ECU also included a fully flexible control of the common-rail (pilot, main and post-injections), the VGT and the EGR, all in the same processor. The current EDC17 (Hammel et al. 2003) model coordinates a bigger number of data from sensors, actuators and models in order to get a full management of the diesel engine.

From all the requirements and purposes of the ECU, the on-board diagnostics (OBD) standards were applied to engines in order to implement safety routines for detecting sensors or engine systems malfunctions (van Basshuysen and Schaefer 2004). Currently, about 40 % of the automotive electronics are devoted to OBD functions. The gradual introduction of the AT systems in diesel engines, especially with the current trade-off between selecting SCR (more complex due to the urea dosing but effective) or LNT (more simple but less effective), is requiring a renewal of the current diesel air path control (Bickerstaffe 2009), and new functions and managing strategies are being developed; see for instance (Yang et al. 2009) who propose an independent control unit for the AT line separated from the main ECU.

2.2.4.1 Summary on the Information Required for a Global Diesel Management

The commercial diesel engine management focuses on references measured upstream of the combustion chamber, with no direct feedback from the exhaust line. This makes that emissions are controlled in open loop and possible changes and the ageing are not taken into account. The sensors set used in the automotive in the beginning of 00s (Fleming 2001) had hardly changed until the end of the last decade. A major factor is the unnecessary of increasing the number and the type of sensors due to the current control logic. Furthermore, other important factor is the low cost of pressure, temperature and rotational motion sensors (bellow 3$ per unit) against the higher cost (minimum >10–20 $) of more sophisticated sensors, such as NO_x or in-cylinder pressure ones [references obtained from (Fleming 2001) and conversations with engine manufacturers].

Table 2.2 summarises the usual inputs and sensors that are installed in current diesel engines, while indicating possible alternatives in a short future (Turner 2009), which are essentially driven by the emission standards (the need of ATs control and diagnosis). It is rather complicated establishing an universal sensor set and some discrepancies can be found, e.g. GM is starting to include p_{cyl} sensors in engines and some AUDI and BMW vehicles already install NO_x sensors before and/or after the SCR system.

Alternatively, model based strategies present different advantages: models can be used for engine calibration saving costs (Arsie et al. 2011); models can easily be linked in control and diagnosis algorithms as they present faster responses than sensors; and the fusion of sensors and models permits to estimate variables or update parameters on real time solving the drift and ageing and improving the dynamic responses of the original signals.

This section has emphasised the importance of the information about engine states and how to correctly retrieve them on-board. Next section deals with the topic of exhaust gas concentration estimation in CI engines, stressing the variables NO_x and λ.

2.3 Dynamic Exhaust Gas Concentration Estimation

Previous section has made a review on the different technologies in order to manage the diesel engine and has underlined the need of reliable methods for estimating variables online. Test-bed measurement systems are usual for testing and research but these present in general deficient dynamic responses and overall limited on-board

Table 2.2 Summary of the on-board sensors and control inputs used in a commercial LD diesel engine for control and feedback, and split between the main engine subsystems

System	Inputs	Standard sensors	Expected sensors
Injection	u_{prail} $[u_{soi}, u_{id}]_{pilot,main,post}$	n, p_{rail}	$p_{cyl}{}^{*}$ $fuel\ quality$
Air path	u_{vgt}, u_{hp}, u_{lp} u_{it}, u_{bp}	$p_{boost}, \dot{m}_a, T_{cool}$ $T_{boost}, p_{amb}, T_{amb}$	$\dot{m}_{egr}, T_{egr}$ $n_{vgt}, \lambda_{int}{}^{*}$
AT	$[u_{soi}, u_{id}]_{post}, u_{urea}$		$NO_x{}^{*}$, soot, $\lambda_{exh}{}^{*}$

The inputs column defines the main actuator signals in the diesel engine. The remaining columns show the variables that are measured by sensors, differentiating between standard for EURO 5 cars and the possible expectations in a short future. Admittedly, some discrepancies can exist between this distinction depending on the considered car and sector (luxury, utility, etc.). The symbol * represents sensors that can be found in some actual engine models but with less probability. The injection settings are defined by the rail pressure control $u_{p_{rail}}$, the injection timing command (u_{soi}) and the injection duration u_{id}, and these settings can be specified for pilot injections main injections and possible post injections (variable $u_{t_{mi}}$ makes reference to the duration of the main injection only). A detailed summary on the symbols and acronyms used in this work appears in the beginning of this document. The interested reader is also referred to Turner (2009) for finding basic comprehensive explanations about the principles of measurement of the main automotive sensors

capabilities. However, in the last years, and forced by the stringent emission laws, developments of on-board gas concentration sensors have proliferated and some of them are a reality today, such as NO_X or wide-band lambda sensors. Furthermore, COMs and data fusion (DF) techniques that combine different information sources can also be used for real-time (RT) estimations.

Current section gives a review on the different methods for the estimation of the exhaust gas concentration in CI engines, emphasising the on-board methods. Therefore, in addition to being feasible from the point of view of cost and engine implementation, the accuracy and dynamic responses of these estimators should be issued. Hereinafter the following three main possibilities are discussed:

- sensors,
- models and/or virtual sensors, and
- DF techniques for combining different information sources.

2.3.1 Sensors

A subsequent division is made between test bed measurements systems and on-board gas concentration sensors, such as exhaust gas oxygen (EGO), universal exhaust gas oxygen (UEGO) sensors, fast soot sensors and finally innovative solution for monitoring e.g. NO_X by using the substrate materials of the AT systems.

2.3.1.1 Test Bed Sensors

Some of the most common test bed sensors are the following: gas analysers (Horiba 2001) for different gas concentrations; opacimeters based on Beer-Lambert law (Lapuerta et al. 2005) and smoke meters for soot; and gravimetric devices or spectrometers for PM (Mamakos et al. 2013), among others. But these are expensive, require from specific installations and their dynamic responses are questionable. Nevertheless, there have been some developments for fast response emissions analysers for NO_X, CO_2 or particles and portable emission measurement systems (PEMS) (Franco et al. 2013) are being generalised for on-board inspections but with limited dynamic accuracy (Mrosek et al. 2011) and packaging. Finally, standard procedures for measuring emissions in test-cells can show representative errors in the dynamic cycles with respect to the actual transient emissions (Mamakos et al. 2013; Viskup et al. 2011) and this should be taken into account when processing the data (Geivanidis and Samaras 2008; Chan et al. 1997; Arrègle et al. 2006b; Bedick et al. 2009) (see Sect. 2.3.2.3).

2.3.1.2 On-board Gas Concentration Sensors

Oxygen Sensors

In 1968, the introduction of EGO sensors (Dueker et al. 1975) was the key for the implantation of the primary three way catalyst (TWC) control in SI engines. These sensors offered a binary lean/rich resolution for measuring oxygen concentration and thus the equivalent air-to-fuel ratio or lambda, which represents the excess air factor with respect to the stoichiometric air-to-fuel ratio (approximately 14.5 in diesel engines)

$$\lambda = \frac{\dot{m}_a}{\dot{m}_f}\frac{1}{14.5} \tag{2.4}$$

In the end of the 90s, planar UEGO sensors were implemented for SI and CI engines. It is common to use directly the term lambda probe for referring to these sensors. The term lambda is due to the characteristic shape of the signal output in narrow-band lambda sensors, which is similar to the λ letter from the Greek alphabet (see Fig. 2.3). Furthermore, these are also known as switching-type sensors. UEGO sensors present full and linear resolution on oxygen over a wide range, hence the term wide-band. UEGO and EGO response characteristics are shown in Fig. 2.3.

λ measurement in CI engines is used for correcting injector drift. In the case of low temperature combustion (LTC) processes (Lu et al. 2011), intake charge composition, and hence λ is important for the combustion stability and control (Chiang et al. 2007; Agrell et al. 2005). Some studies about wide-band lambda sensors performance can be found in (Regitz and Collings 2008b; Klett et al. 2005).

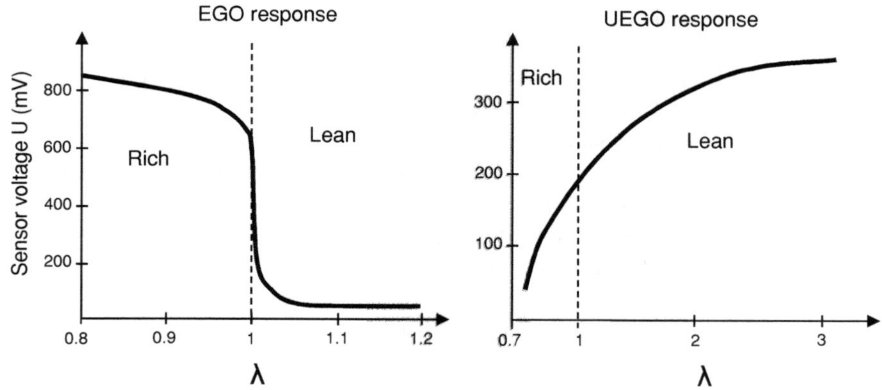

Fig. 2.3 Characteristic outputs of oxygen lambda sensors. *Left* Narrow-band lambda or EGO sensors. *Right* Wide-band lambda or UEGO sensors

NO_x-ZrO_2 Sensors

The development of on-board NO_x sensors has been driven by those which are based on ZrO_2 layers (Kato et al. 1996; Nakanouchi et al. 1996) and able to measure gas at wet condition (i.e. without removing exhaust gas water steam content as gas analysers do). These sensors have suffered an evolution over the last 15 years (Zhuiykov and Miura 2007) and now are manufactured using the planar ZrO_2 multilayer technology (Moos 2005), which combines thick film screen printing and ceramic tape casting (Riegel et al. 2002). Modern versions offer reduced warmup time, smaller size, lower weight and cost-effective production, which encourage their implementation on commercial engines. This kind of sensors simultaneously provides a measurement of the relative air-to-fuel ratio (λ) and NO_x concentration and must play a major role in the SCR control and diagnosis (Hofmann et al. 2004; Hsieh and Wang 2011; Künkel 2001; van Nieuwstadt and Upadhyay 2005).

The working principle is well explained in Riegel et al. (2002) and layout is schematised in Fig. 2.4. This sensor presents two cavities with membranes for measuring oxygen and NO_x, respectively. In a first outer cavity, an electrochemical pump adjusts the oxygen concentration from the diffusing gas to a predefined value (pump intensity I_{p1}) by reducing O_2 to O^{2-} with an electrode (usually Pt), thus providing a linear amperometric measurement of λ. At the same time, a faster binary output is provided, differentiating between rich and lean conditions in the exhaust gases (similar to EGO sensor output). The measurement principle of this first cavity is similar to that of a wide-band UEGO sensor. In addition, NO is oxidised to NO_2, and then the NO_2 is diffused to a second inner cavity, where the oxygen produced through the dissociation of the NO_2 is pumped out in a similar way by means of a second electrochemical pump (I_{p2}). The output of this pump is proportional to the NO_x concentration in the exhaust gases. A heater is necessary for keeping the sensor at high temperatures, where the device gets good resolution, i.e. the sensor needs

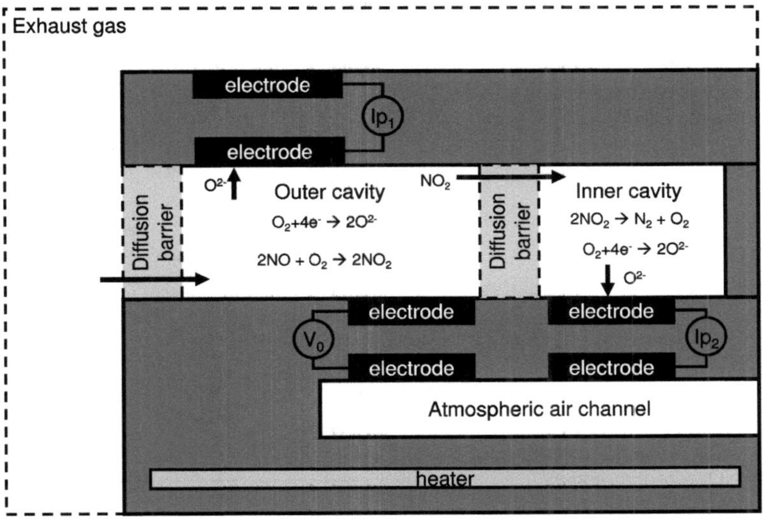

Fig. 2.4 Schematic representation of a NO_X–ZrO_2 sensor, where the main reactions in the chambers and diffusion barriers are shown

some lag for starting to measure when switching on (the same occurs for UEGO sensors). Therefore, these sensors provide three signal outputs: a fast binary λ (around 100 ms), lean or rich, suitable e.g. for TWC diagnosis, a slower (around 500 ms of response time) full resolution λ and NO_X (with a response around 750 ms).

NO_X and UEGO Sensor Limitations

Because of the importance of NO_X and UEGO sensors in the present work, response limitations of these sensors are underlined. Several studies have evaluated the accuracy and response times of real time NO_X sensor measurements for different applications, such as Smith (2000), Manchur and Checkel (2005) and Galindo et al. (2011).

Conventional UEGO sensors exhibit fast response times around 70 ms with sufficient accuracy (Schilling et al. 2008). This can be improved, as proposed in Alberer and del Re (2009a), moving the O_2 sensor upstream of the turbine and using a Kalman filter (1960) for taking into account pressure effect on the output signal. If the NO_X sensor is placed downstream of the AT systems, its response is affected by a considerable transport delay and filtering. Figure 2.5 compares λ^{-1} signals from a UEGO sensor located upstream of the AT systems, and from a NO_X sensor located downstream of an AT line including DOC, DPF and SCR in a turbocharged diesel engine. The NO_X sensor signal is significantly slower and more filtered than that of a UEGO sensor, although the steady-state accuracy seems to be sufficiently accurate in both devices at a glance. Due to the measurement principles of both UEGO and

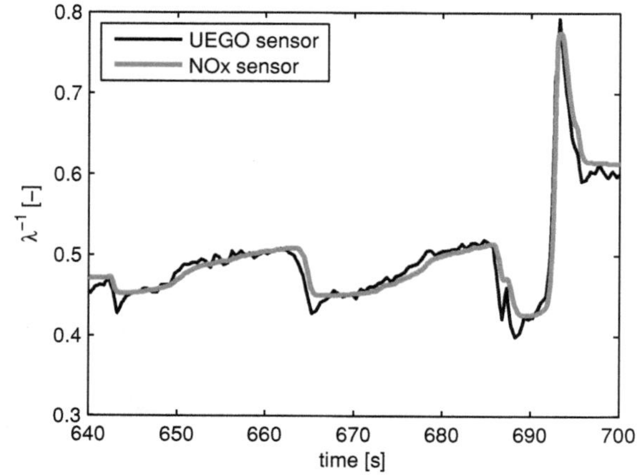

Fig. 2.5 λ^{-1} measurement from a UEGO sensor located at the turbine outlet and from a NO_x sensor downstream the after-treatment systems

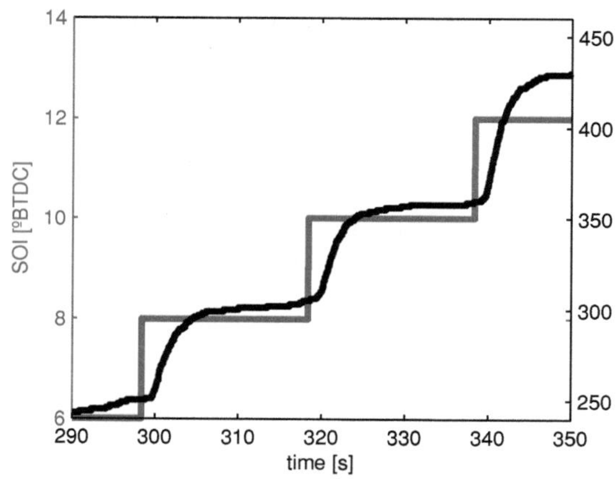

Fig. 2.6 NO_x sensor response to SOI steps (*black-right axis*). NO_x presents a clear delay and filtering with respect to the SOI signal (*gray-left axis*). The dynamic response could be fitted with a first order delayed discrete filter. SOI units are crank angle degrees before the top dead centre (°BTDC), while NO_x is measured in ppm. The delay is in the order of 1 s while the response time is around 0.75 s

NO_x sensors, the pressure in the runners also compromises the steady-state accuracy of the sensor.

Concerning NO_x output from NO_x sensors, the delay is variable, depends on operating conditions, and is in the order of 1 s. This delay is caused for both transport and physical motives, which are difficult to model and can vary with the engine operating conditions. This issue makes difficult to use the raw sensor signals for RT critical functions. For illustrating sensor response, see Fig. 2.6: the actual NO_x is expected to respond instantaneously when performing start of injection (SOI) steps where the delay and response time in the sensor are attributed to the sensor. Several studies have evaluated the accuracy and response times of RT NO_x sensor measurements (Smith 2000; Manchur and Checkel 2005; Mrosek et al. 2011).

Finally, since sensors are subjected to unit-to-unit manufacturing discrepancies, and can be affected by significant drift during their lifespan, methods for the online characterisation of sensors would improve the application of these for automotive control purposes. Usual calibration methods consist of a specific test rig which generates known composition synthesis gas which is used for the static and dynamic calibration of the sensor, while high speed valves are needed for dynamic calibration (Regitz and Collings 2008a); or some solutions based on openly mounted sensors with valves allocated before the sensor, where gas composition is changed within milliseconds (Tobias et al. 1999). However, these calibration methods are restricted to laboratory use and cannot be performed during the operation phase on the engine. Chapter 3 presents a contribution for the online characterisation of NO_x output by means of SOI steps (Galindo et al. 2011); furthermore, λ^{-1} output from wide-band lambda sensors may be characterised by means of load steps.

Fast Soot Sensors

According to (Kasper 2004), particles with lower diameters affect more to alveolar deposition after healthy tests with persons, which underline the risk of underestimating the effect of the residual soot downstream of the DPF. This carries out a sensitive problem for sensors: in addition to robustness, packaging and dynamic limitations, on-board soot sensors must measure with high resolution at low-soot values (this sensitive problem is also linked with NO_x sensors). Another problem with soot sensors is the selection of an appropriate metric as different metrics can be used depending on the physical measuring principle: particle number, total mass, sizes or light absorbing properties (opacity). A common solution for fast soot sensors, described in Warey and Hall (2005), is measuring the electrical conductivity of the carbonaceous fraction of the PM, which acts as a resistance between two electrodes located in the sensor. One of them is feed with a high voltage DC, while the current generated in the other is transformed to voltage, conditioned and then acquired for being converted into a soot estimation. This layout is similar to other fast soot sensors, see e.g. Steppan et al. (2011) and Nelson (2011). Even though these sensors have not reached the maturity of UEGO or NO_x sensors, some companies have already introduced them in the market or are targeted for start of production in 2013.

Sensors in Development, Other Alternatives

A relevant problem of fast NO_x–ZrO_2 sensors is the low sensitivity when the NO_x concentration is small, i.e. downstream of SCR or LNT systems; and the related cross-sensitivity to ammonia. In that field, it can be underlined the work by Moos (Moos et al. 2008; Moos and Schönauer 2008; Moos 2010; Geupel et al. 2011; Schönauer et al. 2011; Groß et al. 2012).

An alternative is using the catalysts and/or substrate materials of the AT devices for determining the NO_x, soot or ammonia by means of measuring the impedance variance of the material. This allows to profit the proper AT device for measuring, developing an integrated device. These are often called in situ monitoring devices.

Moos (2010) show results for in situ sensors in order to detect loadings of oxygen in TWC, NO_x in LNT, NH_3 in SCR-catalyst and soot in DPF, among others. The principal advantage of the approach is the simple and inexpensive setup required as the devices themselves are used as sensors.

Concretely for NO_x sensoring, see Geupel et al. (2011) and Groß et al. (2012) who extend the use of integrating-type NO_x sensors for measuring instantaneous NO_x concentration, especially in the low ppm range (where NO_x concentration is low), incorporating temperature based corrections in the measurements. Groß et al. 2012 also present results with respect to NO/NO_2 differentiation and cross-sensitivities with other gas components. Other alternatives for innovative sensors could be using electromagnetic waves (Moos 2010) or radio-frequency signals which coincide with the oxidation/reduction rate of a TWC for determining oxygen loading without using lambda sensors (Moos et al. 2008).

NH_3 on-board sensors (Moos and Schönauer 2008) are also being developed for closed loop urea dosing control, based on the same principles of the wide-band lambda sensors: the mixed-potential principle between electrodes, separated by a solid electrolyte, where an electric force is created. Delphi has already marketed a commercial version where the reference electrode is in direct contact with exhaust gas (Wang et al. 2008). Schönauer et al. (2011) show results with mixed potential ammonia sensors and suggest some improvements in the technology.

2.3.2 Models and Virtual Sensors

Modelling or virtual sensing[2] is an important tool in the design, analysis and control of IC engines. It is not in vain that model-based strategies in ECU have each time higher responsibilities, overall in control and OBD functions (Stewart et al. 2010). Furthermore, models are quite helpful in the analysis and design phases: hardware and software in the loop (HIL and SIL) are usual because of the reduced cost and time-to-implement with respect to a pure experimental approach. However, the accuracy, reliability, computational cost and time effort of models depend highly on the structure, data used for calibration and the dynamic conditions. A classification on models for automotive is made in Guardiola et al. (2012) and these are compared in Table 2.3:

- Mean value models (MV), which average the physical quantities over a time range, neglecting in-cycle variations. They can be further subdivided in data driven models (DDM) and physical mean value engine models (MVEM).
- Emptying-and-filling models (E&FM), based on first-principle equations and with certain capability for estimating in-cycle variations. They are usually crank-angle solved.

[2] The term virtual sensing is often used for RT on-board models for estimating variables that cannot be measured easily or their measurements are not reliable, in contrast with using physical sensors.

Table 2.3 Summary of the characteristics of different type of models used in the automotive, reproduced from Guardiola et al. (2012)

	DDM	MVEM	E and FM	WAM	CFD
Spatial resolution	0D	0D	0D	1D	3D
Time resolution	MV	MV	CAS	CAS	CAS
Typical acq. rates	0.5–100 Hz	1–200 Hz	>1 MHz	>10 MHz	1–100 MHz
Physical description	No/Very low	Low	Low	Medium	Complex
Prediction capabilities	No	Low	Medium	Medium	High
Computational cost	Very low	Low	Medium	High	Very high
Main application	Control	Control	CAE	CAE	CAE

MV mean value, *acq.* acquisition, *CAS* crank angle solved, *CAE* computer aided-engineering

- Wave Action Models (WAM), able to quantify variables in 1D space and based on an Euler equation balance. These allow to represent wave effects (useful e.g. for valve lifting tuning).
- Computational fluid dynamics models (CFD), based on a detailed geometrical description of the engine volumes and the application of the Navier-Stokes equations. Due to the 3D nature of this approach, this tool is effective for simulating mixing models, i.e. EGR distribution around the cylinders or injection spray models. However, CFD characteristic times of computation can range from seconds to days, preventing its use for RT purposes.

In order to design control oriented models (COMs), MVEM and DDM are the predominant approaches (Calendini and Breuer 2010), basically for the relatively simple computation and the good accuracy achieved, especially when the training data set is appropriate, getting indeed higher accuracy than more complex models. MVEM and DDM rely on two main hypothesis: variables are averaged over the time, with computation (and acquisition) rates about 50 Hz,[3] and the number of states that these models can manage are reduced due to the limited computational resources available (mainly the ECU limitations). A straightforward classification of COMs is:

- Physical models,
- gray box models, which combine first principle equations with parameters, curves and maps to be fitted with engine data, and
- black box models, which are fitted only by data and with no physical based structure.

These approaches present advantages and drawbacks. In the following, COMs for λ and NO_x are discussed, as being core variables of the present work. The interested reader can find a review on proposed models for these variables in the dissertation by (Schilling 2008) with application to failure detection and isolation (FDI).

[3] For an engine that is spinning at 3,000 rpm, that is equivalent to a rate of 1–100 Hz. A typical frequency for getting changes inside the combustion chamber could be in the order of 1° if talking about crank angle, and this makes a required frequency (at 3,000 rpm) of $50 \times 360 = 18$ MHz, quite higher than the typical rates used in MV models.

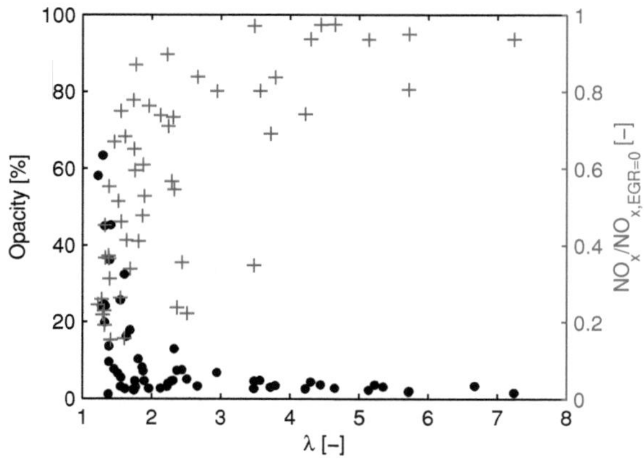

Fig. 2.7 Opacity (*left axis and black points*) and nominal NO_x (*gray* +) emissions depending on λ for nominal tests in the TCCI engine used in Chap. 3. NO_x is normalised with the maximum NO_x value for each operating point, which corresponds to $u_{egr} = 100$ (fully closed) for each pair of $[n, m_f]$

2.3.2.1 λ Models

λ, although not measuring emissions itself, is a key variable in order to control emissions in CI engines. Figure 2.7 shows the opacity and NO_x with respect to λ showing the clear opposite trade-off between both variables. λ can be directly estimated in the basis of $\dot{m}_a$ and $\dot{m}_f$, see (2.4). Both signals can be measured from the ECU or estimated by a model. In the first case, $\dot{m}_a$ is measured by a sensor and $\dot{m}_f$, inferred by a the p_{rail} and $u_{t_{mi}}$ (2.1a). This computation is simple and no specific model is necessary.

Alternatively, MVEMs can be proposed for calculating λ, which will be reduced to estimate the air mass flow at intake as m_f is often considered as an input for control models (Guardiola et al. 2013c). Basic literature for principle equations of CI engines is in Heywood (1988), Guzzella and Amstutz (1998) and Guzzella and Onder (2004) while Wahlström (2009) develops and tune a MVEM for control of EGR and VGT in diesel engines. In addition, other model types with higher or lower ability to compute on RT can be proposed, Tunestsl and Hedrick (2003) estimate λ using net heat release data and (Arsie et al. 2006) use neural networks (NN) (with SI application), among other examples (Nyberg and Stutte 2004; Cesario et al. 2005).

2.3.2.2 NO_x Models

The case of emissions, such as NO_x, is more complex than for λ. NO_x can be inferred by using predictive models that can be implemented on-engine for RT purposes such as engine or AT control and diagnosis. Furthermore, such models can be useful in an off-line basis for NO_x simulation and prototype design. The physical based approach presents difficulties to be applied on RT but can be the basis for grey box approaches (Arsie et al. 2003). First, some words on NO_x physics.

NO_x Physical-Based Models

The term NO_x includes all nitrogen oxides but the nitric oxide (NO) is the predominant in diesel engines (Heywood 1988). NO_x formation is affected by three different mechanisms: thermal, prompt and from fuel-bound nitrogen (Arrègle et al. 2010). The thermal mechanism is the most important in diesel engines where high temperatures benefit reaction of N and O_2 from air producing NO_x. NO_x formation physics in combustion and explosion processes were modelled by Zeldovych (1946) in 1946, and formulated for IC engines by Lavoie et al. (1970) in 1970 with the well-known extended-Zeldovych mechanism

$$\begin{aligned} N + NO &\rightleftarrows NO_2 \\ N + O_2 &\rightleftarrows NO + O \\ N + OH &\rightleftarrows NO + H \end{aligned} \tag{2.5}$$

According to Heywood (1988), the typical characteristic times of the NO formation in diesel engines combustion is in the order of seconds and thus under the hypothesis of equilibrium of certain species, the $d\mathrm{NO}/dt$ can be fitted with the initial NO formation rate by the Arrhenius equation

$$\frac{d\mathrm{NO}}{dt} = \frac{k_1}{T^{0.5}} e^{-k_2/T} [\mathrm{O_2}]^{0.5} [\mathrm{N}] \tag{2.6}$$

The strong dependency with the temperature T is clear: when T increases, NO (and thus NO_x) increases exponentially. The other mechanisms can be relevant in some specific conditions such as LTC (Andersson 2006; Desantes et al. 2012).

Cylinder conditions such as temperature, pressure and oxygen concentration (Brand 2005; Luján et al. 2008; Timoney et al. 2005) are the most important variables for determining NO_x concentration.

Because of the cylinder severe conditions, price and signals problems have prevented to use in-cylinder sensors in commercial vehicles. Although in-cylinder pressure cost is one of the main factors burdening its application for automotive engines, the continuous improvements of pressure sensors and its applications are justifying its nearby implementation (Powell 1993). The case of the temperature sensors is more complicated (Fleming 2001), and their use is not foreseen in applications. Therefore, the solution for estimating the in-cylinder conditions on RT is then using virtual sensors.

Physical models for NO_x estimation are often based in the heat release, using pressure sensor signal for estimating flame temperature (T_f) in the cylinder. The problem is not trivial and a multi-zone discretisation is advisable. Furthermore, the residual gases (internal EGR) affect the process and these are not always easy to estimate. An usual solution to the problem is applying mass and energy conservation equations to each zone (walls, injector neighbourhood, etc.) in conjunction with heat transfer equations, among them it stands out the heat transfer to cylinder wall (Payri et al. 2005), which can be approached by using Woschni equation (Woschni 1967).

Finally, NO_x emissions for both Diesel and gasoline engines are calculated by using the extended Zeldovych mechanism. Good examples for estimating NO_x by using heat release are Arrègle et al. (2006a), Payri et al. (2011a), Guardiola et al. (2011) and Westlund et al. (2008). In the case of SI engines, some authors use the ionisation current on the spark for approaching the pressure (Andersson and Eriksson 2002, 2009), but this is not available on the diesel engine. Nevertheless, the accuracy of the prediction is not as satisfactory as expected, and the drift cannot be eliminated.

There is an open discussion about using time or crank angle based models for the NO_x prediction. For the case of pressure based models, crank angle sampling seems more logical as volume can be easily linked with pressure trace. However, these models require heavy calculations and big memory resources. The time required for completing one engine cycle is often bigger than the characteristic time of the engine. In order to overcome this limitation some authors have proposed simplifications. Guardiola et al. (2011) develop a semi-physical discrete event model based on the heat release calculation but considering only one zone as the main contributor to the NO_x formation; and the process is supposed adiabatic, approaching T_f with the adiabatic temperature of the process. This approach requires specific corrections, especially when the combustion temperature is low. Other examples are Westlund (2011) who present a fast physical model for NO_x and soot, or Arsie et al. (2003) who present a hierarchical model structure for engine control design with different models and layers, ranging from physical based to mean value approaches.

Control-Oriented Models for NO_x

Literature about COMs for NO_x is extensive. Schilling (2008) gives a short review on different COMs for NO_x and presents a NO_x virtual sensor (see also references therein). His work is based on deriving maps from complex models (Schilling et al. 2006, 2008) or engine data (Lughofer et al. 2011; Desantes et al. 2011), building a model that is easy to integrate into the engine ECU, but with limited extrapolation capabilities. In commercial ECUs the prevailing approach is to use look-up tables to model nonlinear and operating point dependent behaviours because of the simple programming in spite of the intensive tuning effort (Guardiola et al. 2013a); even though other possibilities can be commented. Winkler-Ebner et al. (2010) design a virtual NO_x sensor for SCR control and diagnosis. Black box models rely on system identification (Karlsson et al. 2010) but are often operating point dependent and its adaptation is not an easy task. Hirsch et al. (2008) present a gray box model for NO_x and PM. Other non-linear approaches are Takagi-Sugeno fuzzy models (Takagi and Sugeno 1985; Lughofer et al. 2011), Hammerstein-Wiener (HW) (Falck et al. 2012), or NN (Yen and Michel 1991; Alonso et al. 2007; Maaß et al. 2009; Arsie et al. 2010).

2.3.2.3 Identification of Sensor and Physical Dynamics

Dynamic models describe the evolution of a state vector x along the time t and subjected to a specific set of inputs u

$$\dot{x} = f(x, u, t) \quad x \in \mathbb{R}^{n_x}, \quad u \in \mathbb{R}^{n_u} \tag{2.7a}$$

$$y = g(x, u, t) \tag{2.7b}$$

The identification is the technique of designing and tuning the dynamic models by means of a data set $\{u, y\}_{t_1}^{t_{end}}$ over a certain time t_{end}; see (Johansson 1993; Ljung 1999) for comprehensives references on system identification. A common procedure for identification is defining different black box linear model structures (AR, ARX, ARMAX, etc.) based on polynomial relationships with a certain order and solving the linear regression problem, usually by means of a least squares (LS) formulation. In addition to the model parameters, the selection of an appropriate order of the system is key for an adequate solution. If the system is non-linear, specific structures should be utilised, e.g. piece-wise affine (PWA), linear parameter varying (LPV) or HW models, among others.

Sensors Identification for On-board Strategies

Most of the dynamic models with physical meaning used in engineering, as for instance sensor models, respond to n-order linear filters with a certain delay τ (Ogata 2001), at least in the selected areas of performance. In the case of exhaust gas concentration sensors, 1st order linear models can be sufficient for representing the sensor response in the s-domain (Galindo et al. 2011; Mrosek et al. 2011)

$$G(s) = e^{-s\tau} \frac{k}{1 + Ts} \tag{2.8}$$

where T is the response time and k is the gain. Parameters can be fitted by LS. However, actual sensor responses are non-linear and parameters usually vary, especially the delay or dead-time τ, due to the hardware itself, the engine operating conditions and ageing. These variations can be modelled by scheduling strategies, where parameters are stored in look-up tables, or functions (Trimboli et al. 2012) typically depending on n and m_a.

Another possibility is using model structures with physical insight: e.g. Wang (2007) identifies the oxygen sensors dynamics by means of a physical based model and Zhuiykov (2010) describes and models the electrochemistry of ZrO_2 gas sensors. Adaptive filtering can also be used for taking into account the ageing and dispersion. Furthermore, if the signal processing is made off-line, non-causal deconvolution techniques may also be used, see e.g. Henningsson (2012).

DDMs are often based on static relationships such as look-up tables and algebraic equations. In addition, the dynamics are often modelled by linear filters and the correct identification is critical for the adaptive filtering algorithms presented hereinafter.

2.3.3 Adaptive Filtering

Sensors present problems when using the raw output signal in RT functions, in particular the delay from engine to sensor and the response time of the sensor. With respect to models, two problems can be underlined when working with COMs. On one hand, the model accuracy is driven by the collection of the appropriate data and calibration of all the parameters. This is a hard and time consuming task. In fact, ECU has a big number of maps and parameters for engine and vehicle management. On the other hand, independently of how well the model has been calibrated there is inevitably a drift between the system and the model as the surrounding conditions changes and the engine ages. DDMs are highly sensitive to the calibration data set and will have problems with the ageing, manufacturing discrepancies, slowly varying parameters and other non-modelled variables. These affects also to dynamic models identified for both sensor and physics filtering.

2.3.3.1 Data Fusion

Data fusion (DF) utilises information from different sources for providing a better estimation of a given variable. These information sources can come from sensors, models and/or other DF algorithms. DF has been used for many years in different engineering fields, such as e.g. inertial navigation of satellites and missiles (Wagner and Wieneke 2003) or automotive applications in intelligent transportation systems (Faouzi et al. 2011; Stiller et al. 2011) related with traffic and driver/road assistance, among others. Different DF methods can be used as (Faouzi et al. 2011) propose: statistical, ranging from the simplest arithmetic mean to weighted combinations or data mining techniques (Henningsson 2012); probabilistic, such as Bayesian approaches, maximum likelihoods methods and Kalman filter (KF); and artificial intelligence including NN or genetic algorithms. A more exhaustive classification and discussion on challenging problems of multisensor DF is in (Khaleghi et al. 2013). From DF methods, those which are applied adaptively for bounding error of parameters, signals and parameters can be included as methods for adaptive filtering, fusion or learning (Mehra 1970).

The present work centres on the use of Kalman filter (KF) (1960) based tools in order to observe exhaust gas concentration variables in CI engines (Guardiola et al. 2013b, c). The KF is probably the most extended adaptive filtering algorithm used in engineering (Gao and Harris 2002) and provides a systematic and simple way of manipulating linear dynamic models and engine variables by supposing a priori knowledge of the measurement and noise statistics, modelled by Gaussian distributions. This property allows to track the estimation error by means of a covariance matrix **P** that also tracks the ageing of the states, i.e. if the error in a given state is foreseen to be higher or lower depending on when was updated the last time and how much. The KF minimises this expected error by solving an iterative Riccati matrix equation and setting out a Kalman gain (K) for correction. The linear structure of the filter makes it appropriate to ECU-oriented approaches, even though simplifications shall be considered when manipulating large state vectors. Furthermore, depending on the available data, different fusion structures can be programmed as Gao and Harris (2002) show.

Other filtering alternatives, such as RLS (recursive least squares) or proportional methods can also be applied and in some cases lead to similar solutions. Input observers by means of measurements of the states and outputs can also be addressed for estimating engine variables (Chadli et al. 2009; Stotsky and Kolmanovsky 2002). The interested reader is referred to Simon (2006) and Höckerdal (2011), the former for finding a complete and broader view on optimal state estimation and KF-based algorithms for engineering applications, and the latter for a key precedent and motivation for the current work.

In the next, and due to the importance of the KF in this work, main equations are recalled although the reader familiarised with control engineering could skip this paragraph.

2.3.3.2 Kalman Filter

In the setting the data are assumed to be generated by the following discrete time system

$$x_k = f(x_{k-1}, u_k) + w_k \tag{2.9a}$$

$$y_k = g(x_k, u_k) + v_k \tag{2.9b}$$

where $x_k \in \mathbb{R}^{n_x}$ represents the state vector, $u_k \in \mathbb{R}^{n_u}$ the input vector, $y_k \in \mathbb{R}^{n_y}$ the output vector. If f and/or g are non-linear, a previous linearisation step is required for the filter and then, elements ij of $\mathbf{F_k}$ and $\mathbf{H_k}$ are obtained

$$\mathbf{F_{k,ij}} = \frac{\partial \mathbf{f_i}}{\partial \mathbf{x_j}}|_{\mathbf{x}=\hat{\mathbf{x}}_\mathbf{k}} \quad \mathbf{H_{k,ij}} = \frac{\partial \mathbf{g_i}}{\partial \mathbf{x_j}}|_{\mathbf{x}=\hat{\mathbf{x}}_\mathbf{k}} \tag{2.10}$$

being **F** the linearised process matrix and **H** the linearised output matrix. From now, the discussion is valid both for linear and non-linear systems and KF will be used referring to Kalman filter based methods, including the non-linear Extended Kalman Filter (EKF) (Simon 2006) and the standard one.

Noises in (2.9) $w_k \in \mathbb{R}^{n_x}$ and $v_k \in \mathbb{R}^{n_y}$ are assumed to be independent and both generated by Gaussian distribution with zero mean and covariance matrices $\mathbf{Q_k}$ resp. $\mathbf{R_k}$, defined by

$$E[w_k\, w_k^T] = \mathbf{Q_k} \tag{2.11a}$$
$$E[v_k\, v_k^T] = \mathbf{R_k} \tag{2.11b}$$

In many applications these are often chosen to be constant, i.e. **Q** and **R**, and diagonal.

Then, $\hat{x}_k \in \mathbb{R}^{n_x}$ is the observation of the state vector x_k

$$\hat{x}_{k|k-1} = f(\hat{x}_{k-1}, u_k) \tag{2.12a}$$
$$e_k = y_k - g(\hat{x}_{k|k-1}, u_k) \tag{2.12b}$$
$$\hat{x}_k = \hat{x}_{k|k-1} + K_k\, e_k \tag{2.12c}$$

where Kalman gain K_k is solved by the following iterative equation

$$\mathbf{P_{k|k-1}} = (\mathbf{F_k P_{k-1} F_k^T} + \mathbf{Q}) \tag{2.13a}$$
$$K_k = \mathbf{P_{k|k-1} H_k^T} \left(\mathbf{H_k P_{k|k-1} H_k^T} + \mathbf{R}\right)^{-1} \tag{2.13b}$$
$$\mathbf{P_k} = (\mathbf{I} - K_k \mathbf{H_k})\mathbf{P_{k|k-1}} \tag{2.13c}$$

and matrix $\mathbf{P_k}$ is the covariance matrix of the state estimate error (Ljung 1999)

$$\mathbf{P_k} = E[x_k - \hat{x}_k][x_k - \hat{x}_k]^T. \tag{2.14}$$

2.3.3.3 About the Methods in this Work

From all the possible applications of the KF for the gas concentration estimation, the following are exploited in the present work:

- Dynamic estimation of engine variables by means of fusing a fast reference, often a model, and a slow but accurate sensor, for developing robust virtual sensors or bias tracking algorithms; and
- on-line adaptation of models, especially the online updating of look-up tables as basic structures in COMs.

These algorithms, although applied in the present work to λ and NO_x, can be used for estimating other relevant engine quantities, such as the intake air mass flow or the volumetric efficiency (Höckerdal et al. 2009, 2011), just considering the appropriate state-space model. In the following, the state of the art related with these applications on automotive is reviewed.

2.3.3.4 Adaptive Estimators Based on the KF

Figure 2.8 shows a model signal for λ^{-1}, based on the injected fuel mass $\dot{m}_f$ estimated by the electronic control unit (ECU) and the air mass flow $\dot{m}_a$ determined from a

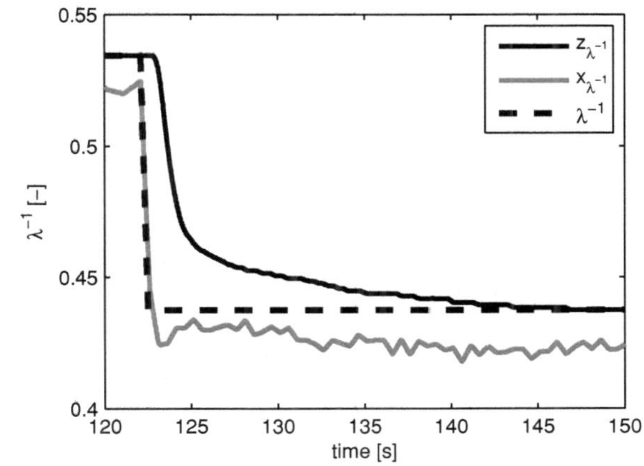

Fig. 2.8 Different λ^{-1} estimations during a load transient in the tested diesel engine (see Sect. 3.2): $z_{\lambda^{-1}}$ is the output of a NO_x sensor located at the turbine outlet, x_{λ}^{-1} is the output of the model presented in Sect. 4.2 and λ^{-1} is the foreseen actual value

hot wire anemometer. Comparing to the sensor, model is faster and non-delayed but presents a bias with respect to the sensor steady-state value.

Algorithms based on bias tracking take a fast model that keeps high frequency components of the considered variable while a slow but steady-state accurate sensor permits to correct model drift. The vector $x_m(t) \in \mathbb{R}^{n_x}$ contains model signals and $z(t) \in \mathbb{R}^{n_z}$ the sensor signals. The drift is modelled with a vector $\theta(t) \in \mathbb{R}^{n_x}$, which contains model biases and varies with time often slowly. If $x_r(t) \in \mathbb{R}^{n_x}$ is the vector containing the actual values of the states, then the true bias θ is

$$\theta(t) = x_r(t) - x_m(t) \tag{2.15}$$

but the problem here is that x_r is usually measured by a sensor z

$$z(t) = g(x_r(t - \tau)) \tag{2.16}$$

and in general will have a certain delay τ. The discussion on identifying the function g and delay τ was made before in Sect. 2.3.2.

In order to design a drift correction algorithm, an augmented state-space model is written. An extra-state is used for tracking the given error (θ), creating a wide state vector x^w with drift vectors and the original state vector $x \in \mathbb{R}^{n_x}$, which contains the estimations of x_m,

$$x^w(t) = \begin{bmatrix} \theta(t) \\ x(t) \end{bmatrix} \tag{2.17}$$

and the state-space model

$$\dot{x}^w = f(x^w, x_m, W, t) \tag{2.18a}$$

$$z = g(x^w, x_m, V, t) \tag{2.18b}$$

is solved by adding noises $W(t) \in \mathbb{R}^{2n_x}$ and $V(t) \in \mathbb{R}^{n_z}$ to the model and sensor output respectively.

This formulation, with different variations, is made by different authors in order to observe the bias between two references, usually a model and a sensor. Alternatively, the bias can be applied to the model and/or sensor by shifting the references when proceeding; for the latter the sensor is not steady-state accurate and a steady-state accurate reference is required (Chen and You 2008).

The classical approach for modelling the drift is supposing a slow variation of the state by the dynamic equation

$$\dot{\theta} = 0 \tag{2.19}$$

and supposing that states noise W is low with respect to output noise V, for having a smooth correction.

There exist different references with similar models as described above for estimating engine variables. The majority of them uses the EKF due to the non-linear equations involved, some references from 2005 are commented below:

- Hsieh and Wang (2011) use an EKF in order to estimate NO_x up, in between and downstream of a two-series SCR configuration and consider the ammonia cross sensitivity of a NO_x sensor located in between.
- Höckerdal et al. (2009) uses an augmented model for observing the sensor bias with an air mass flow sensor application.
- Alberer and del Re (2009a) use a KF for correcting an oxygen measurement made by a UEGO sensor located upstream of the turbine in a TCCI engine, where pressure effects in the sensor are taken into account.
- Polóni et al. (2012) compare two different sensors configuration in a TCCI engine for correcting states in a MVEM of the air path, designing a closed loop MVEM. They use an EKF for tracking the bias on the considered states and then simulate the model with the corrected variables. References included in this work are interesting and show other different observer designs.
- Other examples are Yan and Wang (2012), Surenahalli et al. (2012), Chauvin et al. (2006), Grünbacher et al. (2005), Trimboli et al. (2012), Tschanz et al. (2012), Benaicha et al. (2011), and Zhou et al. (2012) who design observers for different relevant engine quantities, such as intake manifold temperature, soot, individual in-cylinder air to fuel ratio or engine torque among others, with diagnosis or control applications.

Drift Models

However, there are cases where the considered bias, e.g. model bias θ variation does not depend only on time

$$\frac{\mathrm{d}\theta}{\mathrm{d}t} = \frac{\partial\theta}{\partial t} + \frac{\partial\theta}{\partial n}\frac{\mathrm{d}n}{\mathrm{d}t} + \frac{\partial\theta}{\partial m_f}\frac{\mathrm{d}m_f}{\mathrm{d}t} + \cdots \tag{2.20}$$

and although the bias variation associated with the system drift ($\partial\theta/\partial t$) is expected to be slow, the actual variation of the bias may be very fast, due to the ability of the engine of performing fast transition between operating conditions (usually defined by n and m_f). This can be solved by using specific models for the bias, where a possible solution may be using look-up tables.

In the following, the use of adaptive virtual sensors is discussed.

2.3.3.5 Online Adaptation of Models

In addition to the bias and ageing, the basic structure of models lead to errors due to uncertainties and non-considered inputs. Therefore, the model error might be tracked and distributed between the model parameters. This could be made on an offline basis if the variables are stored, and then the model signals and parameters might be updated by means of executing an optimisation method. An alternative is using adaptive filtering for directly updating the model parameters online. Schilling (2008) makes a selection on parameters from a virtual sensor for NO_x and λ; and designs an adaptive virtual sensor for their online adaptation. Another possibility is introduced by Polóni et al. (2012) where a set of individual observers are designed for some states of the model. However, MVEMs often pose a mixed structure with different parameters, curves, tables and dynamic equations. The problem of updating all elements is not straightforward and must be faced carefully. The design of observers for updating parameters and tables can be used for tuning or updating DDMs and MVEMs, although stability and robustness of the solution should be issued. Due to the relevance on the present work, the updating of look-up tables is considered.

Look-up tables[4] allow engineers to easily model systems that present complex expressions by means of mapping outputs with a set of nD heuristic array structures. Look-up tables are used in automotive for different purposes (Vogt et al. 2004): maps of engine parameters as function of the operating-point conditions, often m_f (or torque) and n; in order to provide set-points for the controllers; or for gain scheduling, among others. Furthermore, the manipulation and interpretation of look-up tables is simple and the usual offline procedures for calibration are based on LS methods. Linear interpolation is usual for computing the outputs. Anyhow, the tables are of course subjected to drift and ageing when running online. Adaptive filtering can be used for correcting the drift and/or tuning parameters in an offline basis.

[4] Usual dimensions of look-up tables are 2D, while 1D tables are often named as curves, but in fact they can be considered as 1D look-up tables. Using higher dimensional grids is not usual due to the involved computational burden, excepting specific cases such as for representing different combustion modes.

EKF for Updating Look-up Tables

Höckerdal et al. (2011) propose an EKF for updating table elements where they are treated as states to be observed and updated by measuring the considered output. In this approach, **P** represents the ageing of the states, defining a proper K for updating in such sense that locally inactive elements see how their related variance grows monotonically and active elements see how this is modified depending on the weighted distance to the considered element. The interpolation principle weights the correction and **P** updating: more excited elements will tend to have lower variance than the less excited ones. A bounded limit on **P** should be applied for engineering applications in order to avoid robustness problems.

The first limitation of this algorithm is that although inactive elements do not really affect the updating in a given iteration, EKF manipulates all elements of **P**. The global stability of the method relies on the observability (linked with activeness) of the states and although during each iteration only 4 elements (in a 2D example) are locally observable and the remaining are unobservable, the calculation involves all grid areas. A second limitation is that system matrices vary with time making impossible to derive a steady-state Kalman filter (Payri et al. 2012). This forces to solve the Riccati equation at every instant for inferring K, involving a huge memory and computational resources. By utilising a numerical example, an EKF for updating a look-up table with 20-by-20 elements gives rise to a covariance matrix of 400-by-400 elements and thereby also a significant computational burden when solving the Riccati equations, which grows rapidly with the grid dimensions. Therefore an important discussion of the work is how the KF can be used or modified to get an efficient updating procedure for look-up tables without an important loss of properties.

RLS and Other Approaches

An alternative solution for estimation is the use of LS techniques, see e.g. Peyton Jones and Muske (2009) who use a RLS with a forgetting factor for updating look-up tables. The programming of a RLS is fairly simple and requires that table grid and data is well distributed for avoiding robustness problems (as the authors literally claim). Vogt et al. (2004) present the normalised least mean square (NMLS) method optimised for online-adaptation of tables and propose an interesting simplification on this method by separating non-active and active elements of the considered table, aspect that is also exploited in this work but for EKF-based methods. The results are good with much lower computation involved but with a slower convergence with respect to the standard RLS.

Another interesting contribution due to the simplicity of the formulation is made by Wu (2006), who treats the problem of table updating as a reverse interpolation problem. A proportional weighting is then used for updating all the elements that were involved in the previous interpolation calculation. The author cites it literally as *multiple nodes proportional distribution*. This weighting is calculated only from the inputs and previous outputs. In that way, the method does not take into account any noise in the measurement, and can be considered as sub-optimal if comparing with the KF, but with a much lower computation and memory weight because only the table values related with observable elements are updated.

This work presents two approaches (SKF and SSKF) with some characteristics of the commented above in order to update look-up tables efficiently.

References

Agrell F, Angström HE, Eriksson B, Wikander J, Linderyd J (2005) Transient control of HCCI combustion by aid of variable valve timing through the use of a engine state corrected CA50-controller combined with an in-cylinder state estimator estimating lambda. In: SAE technical paper 2005-01-2128

Alberer D, del Re L (2009a) Fast oxygen based transient diesel engine operation. In: SAE technical paper 2009-01-0622

Alberer D, del Re L (2009b) Optimization of the transient diesel engine operation. In: SAE technical paper 2009-24-0113

Alonso JM, Alvarruiz F, Desantes JM, Hernández L, Hernández V, Moltó G (2007) Combining neural networks and genetic algorithms to predict and reduce diesel engine emissions. IEEE Trans Evol Comput 11(1):46–55

Andersson M (2006) Fast NO_x prediction in diesel engines. PhD Thesis, Lund University

Andersson I, Eriksson L (2009) A parametric model for ionization current in a four stroke SI engine. ASME

Arrègle J, López JJ, Martín J, Mocholí E (2006a) Development of a mixing and combustion zero-dimensional model for diesel engines. In: SAE technical paper 2006-01-1382

Arrègle J, Bermùdez V, Serrano JR, Fuentes E (2006b) Procedure for engine transient cycle emissions testing in real time. Exp Thermal Fluid Sci 30(5):485–496

Arrègle J, López JJ, Guardiola C, Monin C (2010) On board NO_x prediction in diesel engines: a physical approach. In: del Re L et al. (ed) Automotive model predictive control: models, methods and applications. Springer, London. ISBN-1849960704

Arsie I, Pianese C, Rizzo G (2003) An integrated system of models for performance and emissions in SI engines: development and identification. In: SAE technical paper 2003-01-1052

Arsie I, Pianese C, Sorrentino M (2006) A procedure to enhance identification of recurrent neural networks for simulating air-fuel ratio dynamics in SI engines. Eng Appl Artif Intell 19(1):65–77

Arsie I, Pianese C, Sorrentino M (2010) Development of recurrent neural networks for virtual sensing of NO_x emissions in internal combustion engines. SAE Int J Fuels Lubricants 2(2): 354–361

Arsie I, Criscuolo I, Pianese C, De Cesare M (2011) Tuning of the engine control variables of an automotive turbocharged diesel engine via model based optimization. In: SAE technical paper 2011–24-0146

Bedick CR, Clark NN, Zhen F, Atkinson RJ, McKain DL (2009) Testing of a heavy-duty diesel engine schedule for representative measurement of emissions. J Air Waste Manage Assoc 59(8):960–971

Benaicha F, Bencherif K, Sorine M, Vivalda JC (2011) Model based mass soot observer of diesel particle filter. In: IFAC proceedings volumes (IFAC-PapersOnline), vol 18, pp 10647–10652

Bickerstaffe S (2009) No one ideal solution. Automot Eng 34(9):44–46

Bosch R (2011) Automotive handbook. In: Bosch handbooks, 8th edn. Robert Bosch GmbH

Brand D (2005) Control-oriented modeling of NO emissions of SI engines. PhD Thesis, ETH, Zürich

Calendini PO, Breuer S (2010) Mean value engine models applied to control system design and validation. Lecture notes in control and information sciences, vol 402. Springer, London

Carberry B, Grasi G, Guerin S, Jayat F, Konieczny R (2005) Pre-turbocharger catalyst—Fast catalyst light-off evaluation. In: SAE technical paper 2005-01-2142
Cesario N, Di Meglio M, Pirozzi F, Moselli G, Taglialatela F, Carpentieri F (2005) Air/Fuel control system in SI engines based on virtual lambda sensor. In: SAE technical paper 2005-24-058
Chadli M, Akhenakb A, Ragot J, Maquinc D (2009) State and unknown input estimation for discrete time multiple model. J Franklin Inst 346:593–610
Chan SH, Chen XS, Arcoumanis C (1997) Measurement and signal reconstruction of transient nitric oxide emissions in the exhaust of a turbocharged diesel engine. J Dyn Syst Meas Control, Trans ASME 119(4):620–630
Chauvin J, Moulin P, Corde G, Petit N, Rouchon P (2006) Kalman filtering for real-time individual cylinder air fuel ratio observer on a diesel engine test bench. In: Proceedings of the American control conference, pp 1886–1891
Chauvin J, Grondin O, Moulin P (2011) Control oriented model of a variable geometry turbocharger in an engine with two EGR loops. Oil Gas Sci Technol-Revue D IFP Energies Nouvelles 66(4, SI):563–571
Chen TS, You RZ (2008) A novel fault tolerant sensor system for sensor drift compensation. Sens Actuators A: Phys 147:623–632
Chiang CJ, Stefanopoulou AG, Jankovic M (2007) Transitions in homogeneous charge compression ignition engines. IEEE Trans Control Syst Technol 15(3):438–448
Chi J, Da Costa H (2005) Modeling and control of a urea-SCR aftertreatment system. In: SAE technical paper 2005-01-0966
Deng J, Stobart R, Liu C, Winward E (2012) Explicit model predictive control of the diesel engine fuel path. In: SAE technical paper 2012-01-0893
Desantes JM, Luján JM, Guardiola C, Blanco-Rodriguez D (2011) Development of NO_x fast estimate using NO_x sensors. In: EAEC 2011 congress, Valencia
Desantes JM, López JJ, Redón P, Arrègle J (2012) Evaluation of the thermal NO formation mechanism under low-temperature diesel combustion conditions. Int J Engine Res 13(6):531–539
Desantes JM, Luján JM, Pla B, Soler JA (2013) On the combination of high-pressure and low-pressure exhaust gas recirculation loops for improved fuel economy and reduced emissions in high-speed direct-injection engines. Int J Engine Res 14(1):3–11
Dueker H, Friese KH, Haecker WD (1975) Ceramic aspects of the bosch lambda-sensor. In: SAE technical paper 750223
Ebert C, Jones C (2009) Embedded software: facts, figures, and future. IEEE Comput 42(4):42–52
Eriksson L, Andersson I (2002) An analytic model for cylinder pressure in a four stroke SI engine. In: SAE technical paper 2002–01-0371
Falck T, Dreesen P, De Brabanter K, Pelckmans K, De Moor B, Suykens JAK (2012) Least-squares support vector machines for the identification of Wiener-Hammerstein systems. Control Eng Pract 20(11):1165–1174
Faouzi N-E, Leung H, Kurian A (2011) Data fusion in intelligent transportation systems: progress and challenges—A survey. Inf Fusion 12(1):4–10
Fleming WJ (2001) Overview of automotive sensors. IEEE Sens J 1(4):296–308
Franco V, Kousoulidou M, Muntean M, Ntziachristos L, Hausberger S, Dilara P (2013) Road vehicle emission factors development: a review. Atmos Environ 70:84–97
Frobert A, Raux S, Lahougue A, Hamon C, Pajot K, Blanchard G (2012) HC-SCR on silver-based catalyst: from synthetic gas bench to real use. SAE Int J Fuels Lubricants 5(1):389–398
Galindo J, Climent H, Guardiola C, Doménech J (2009) Strategies for improving the mode transition in a sequential parallel turbocharged automotive diesel engine. Int J Automot Technol 10(2): 141–149
Galindo J, Serrano JR, Guardiola C, Blanco-Rodriguez D, Cuadrado IG (2011) An on-engine method for dynamic characterisation of NO_x concentration sensors. Exp Thermal Fluid Sci 35(3): 470–476
Gao JB, Harris CJ (2002) Some remarks on Kalman filters for the multisensor fusion. Inf Fusion 3:191–201

Geivanidis S, Samaras Z (2008) Development of a dynamic model for the reconstruction of tailpipe emissions from measurements on a constant volume sampling dilution system. Meas Sci Technol 19(1).doi:10.1088/0957-0233/19/1/015404

Geupel A, Kubinski DJ, Mulla S, Ballinger TH, Chen H, Visser JH, Moos R (2011) Integrating NO_x sensor for automotive exhausts—A novel concept. Sens Lett 9(1):311–315

Groß A, Beulertz G, Marr I, Kubinski DJ, Visser JH, Moos R (2012) Dual mode NO_x sensor: measuring both the accumulated amount and instantaneous level at low concentrations. Sensors (Basel) 12(3):2831–2850

Grünbacher E, Kefer P, del Re L (2005) Estimation of the mean value engine torque using an extended Kalman filter. In: SAE technical paper 2005–01-0063

Guardiola C, López JJ, Martín J, García-Sarmiento D (2011) Semiempirical in-cylinder pressure based model for NO_x prediction oriented to control applications. Appl Thermal Eng 31(16):3275–3286

Guardiola C, Gil A, Pla B, Piqueras P (2012) Representation limits of mean value engine models. Lecture notes in control and information sciences, vol 418, pp 185–206

Guardiola C, Pla B, Blanco-Rodriguez D, Cabrera P (2013a) A learning algorithm concept for updating look-up tables for automotive applications. Math Comput Model 57(7–8):1979–1989

Guardiola C, Pla B, Blanco-Rodriguez D, Eriksson L (2013b) A computationally efficient Kalman filter based estimator for updating look-up tables applied to NO_x estimation in diesel engines. Control Eng Pract 21(11):1455–1468

Guardiola C, Pla B, Blanco-Rodriguez D, Mazer A, Hayat O (2013c) A bias correction method for fast fuel-to-air ratio estimation in diesel engines. In: Proceedings of the Institution of Mechanical Engineers. Part D: J Automobile Eng 227(8):1099–1111

Guzzella L, Amstutz A (1998) Control of diesel engines. IEEE Control Syst Mag 8:55–71

Guzzella L, Onder CH (2004) Introduction to modeling and control of internal combustion engine systems. Springer, Berlin

Haber B, Wang J (2010) Robust control approach on diesel engines with dual-loop exhaust gas recirculation systems. In: ASME 2010 dynamic systems and control conference, DSCC2010, vol 1, pp 711–718

Hammel C, Jessen H, Boss B, Traub A, Tischer C, Hönninger H (2003) A common software architecture for diesel and gasoline engine control systems of the new generation EDC/ME(D)17. In: SAE technical paper 2003–01-1048

Henningsson M (2012) Data-rich multivariable control of heavy-duty engines. PhD Thesis, Department of Automatic Control, Lund University, Sweden, Mayo 2012

Heywood JB (1988) Internal combustion engine fundamentals. McGraw Hill, New York

Hirsch M, Alberer D, del Re L (2008) Grey-box control oriented emissions models. In: Proceedings of the 17th world congress, The International Federation of Automatic Control, Seoul, Korea, 6–11 July 2008

Höckerdal E (2011) Model error compensation in ODE and DAE estimators with automotive engine applications. PhD Thesis, Linköping University Institute of Technology

Höckerdal E, Frisk E, Eriksson L (2009) Observer design and model augmentation for bias compensation with a truck engine application. Control Eng Pract 17(3):408–417

Höckerdal E, Frisk E, Eriksson L (2011) Ekf-based adaptation of look-up tables with an air mass-flow sensor application. Control Eng Pract 19:442–453

Hofmann L, Rusch K, Fischer S, Lemire B (2004) Onboard emissions monitoring on a HD truck with an SCR system using NO_x sensors. In: SAE technical paper 2004–01-1290

Horiba (2001) Horiba MEXA-7000DEGR instruction manual, Aug 2001

Hsieh M-F, Wang J (2011) Design and experimental validation of an extended Kalman filter-based NO_x concentration estimator in selective catalytic reduction system applications. Control Eng Pract 19(4):346–353

Johansson R (1993) System modeling and identification. Prentice Hall information and system sciences series. Prentice Hall, Englewood Cliffs

Johnson TV (2012) Vehicular emissions in review. In: SAE technical paper 2012–01-0368, vol 5(2)

Kalman RE (1960) A new approach to linear filtering and prediction problems. J Basic Eng 82 (35–45)

Karlsson M, Ekholm K, Strandh P, Tunestål P, Johansson R (2010) Dynamic mapping of diesel engine through system identification. In: Proceedings of the American control conference, Baltimore, MD

Kasper M (2004) The number concentration of non-volatile particles—Design study for an instrument according to the PMP recommendations. In: SAE technical paper 2004–01-0960

Kato N, Nakagaki K, Ina N (1996) Thick film ZrO_2 NO_x sensor. In: SAE technical paper 960334

Khaleghi B, Khamis A, Karray FO, Razavi SN (2013) Multisensor data fusion: a review of the state-of-the-art. Inf Fusion 14(1):28–44

Klett S, Piesche M, Heinzelmann S, Weyl H, Wiedenmann H-M, Schneider U, Diehl L, Neumann H (2005) Numerical and experimental analysis of the momentum and heat transfer in exhaust gas sensors. In: SAE technical paper 2005–01-0037

Künkel C (2001) Catalytic reduction of NO_x on heavy-duty trucks. PhD Thesis, Lund University

Ladommatos N, Abdelhalim S, Zhao H, Hu Z (1996a) The dilution, chemical, and thermal effects of exhaust gas recirculation on diesel engine emissions. In: Part 1: Effect of reducing inlet charge oxygen. SAE technical paper 961165

Ladommatos N, Abdelhalim S, Zhao H, Hu Z (1996b) The dilution, chemical, and thermal effects of exhaust gas recirculation on diesel engine emissions. In: Part 2: Effects of carbon dioxide. SAE technical paper 961167

Ladommatos N, Abdelhalim S, Zhao H, Hu Z (1997a) The dilution, chemical, and thermal effects of exhaust gas recirculation on diesel engine emissions. In: Part 3: Effects of water vapour. SAE technical paper 971659

Ladommatos N, Abdelhalim S, Zhao H, Hu Z (1997b) The dilution, chemical, and thermal effects of exhaust gas recirculation on diesel engine emissions. In: Part 4: Effects of carbon dioxide and water vapour. SAE technical paper 971660

Ladommatos N, Abdelhalim S, Zhao H (2000) The effects of exhaust gas recirculation on diesel combustion and emissions. Int J Engine Res 1(1):107–126

Lapuerta M, Martos FJ, Cárdenas MD (2005) Determination of light extinction efficiency of diesel soot from smoke opacity measurements. Meas Sci Technol 16(10):2048–2055

Lavoie GA, Heywood JB, Keck JC (1970) Experimental and theoretical study of nitric oxide formation in internal combustion engines. Combust Sci Technol 1(4):313–326

Ljung L (1999) System identification: theory for the user. Prentice Hall PTR, Upper Saddle River

Lu X, Han D, Huang Z (2011) Fuel design and management for the control of advanced compression-ignition combustion modes. Prog Energy Combust Sci 37:741–783

Lughofer E, Macián V, Guardiola C, Klement EP (2011) Identifying static and dynamic prediction models for NO_x emissions with evolving fuzzy systems. Appl Soft Comput 11:2487–2500

Luján JM, Pla B, Moroz S, Bourgoin G (2008) Effect of low pressure EGR on gas exchange processes and turbocharging of a HSDI engine. In: THIESEL 2008 conference on thermo- and fluid-dynamic processes in diesel engines

Maaß, Stobart R, Deng J (2009) Diesel engine emissions prediction using parallel neural networks. In: Proceedings of the American control conference, pp 1122–1127

Mamakos A, Bonnel P, Perujo A, Carriero M (2013) Assessment of portable emission measurement systems (PEMS) for heavy-duty diesel engines with respect to particulate matter. J Aerosol Sci 57:54–70

Manchur TB, Checkel MD (2005) Time resolution effects on accuracy of real-time NO_x emissions measurements. In: SAE technical paper 2005–01-0674

Mehra RK (1970) Approaches to adaptive filtering. In: IEEE symposium on adaptive processes, Austin

Millo F, Giacominetto PF, Bernardi MG (2012) Analysis of different exhaust gas recirculation architectures for passenger car diesel engines. Appl Energy 98:79–91

Moos R (2005) A brief overview on automotive exhaust gas sensors based on electroceramics. Int J Appl Ceram Technol 2(5):401–413

Moos R (2010) Catalysts as sensors—A promising novel approach in automotive exhaust gas aftertreatment. Sensors (Basel) 10(7):6773–6787

Moos R, Schönauer D (2008) Recent developments in the field of automotive exhaust gas ammonia sensing. Sens Lett 6(6):821–825

Moos R, Spörl M, Hagen G, Gollwitzer A, Wedemann M, Fischerauer G (2008) TWC: lambda control and OBD without lambda probe—An initial approach. In: SAE technical paper 2008–01-0916

Mrosek M, Sequenz H, Isermann R (2011) Identification of emission measurement dynamics for diesel engines. In: IFAC proceedings volumes (IFAC-PapersOnline), vol 18, pp 11839–11844

Nakanouchi Y, Kurosawa H, Hasei M, Yan Y, Kunimoto A (1996) New type of NOx sensors for automobiles. In: SAE technical paper 961130

Nelson CS (2011) Particulate matter sensor. US patent 8225648

Nyberg M, Stutte T (2004) Model based diagnosis of the air path of an automotive diesel engine. Control Eng Pract 12:513–525

Ogata K (2001) Modern control engineering, 4th edn. Prentice Hall, Englewood Cliffs

Ortner P, del Re L (2007) Predictive control of a diesel engine air path. IEEE Trans Control Syst Technol 15(3):449–456

Park S, Matsumoto T, Oda N (2010) Numerical analysis of turbocharger response delay mechanism. In: SAE technical paper 2010-01-1226

Payri F, Margot X, Gil A, Martín J (2005) Computational study of heat transfer to the walls of a DI diesel engine. In: SAE technical paper 2005–01-0210

Payri F, Olmeda P, Martín J, García A (2011a) A complete 0D thermodynamic predictive model for direct injection diesel engines. Appl Energy 88(88):4632–4641

Payri F, Serrano JR, Piqueras P, García-Afonso O (2011b) Performance analysis of a turbocharged heavy duty diesel engine with a pre-turbo diesel particulate filter configuration. SAE Int J Engines 4(2):2559–2572

Payri F, Guardiola C, Blanco-Rodriguez D, Mazer A, Cornette A (2012) Methodology for design and calibration of a drift compensation method for fuel-to-air ratio estimation. In: SAE technical paper 2012–01-0717

Peyton Jones JC, Muske KR (2009) Identification and adaptation of linear look-up table parameters using an efficient recursive least-squares technique. ISA Trans 48(4):476–483

Pla B (2009) Análisis del Proceso de la Recirculación de los Gases de Escape de Baja Presión en Motores Diesel Sobrealimentados. PhD Thesis, Universitat Politècnica de València, Departamento de Máquinas y Motores Térmicos

Polóni T, Rohal'-Ilkiv B, Alberer D, del Re L, Johansen TA (2012) Comparison of sensor configurations for mass flow estimation of turbocharged diesel engines, vol 418 of Lecture notes in control and information sciences

Powell JD (1993) Engine control using cylinder pressure: past, present, and future. J Dyn Syst Meas Control, Trans ASME 115(2 B):343–350

Regitz S, Collings N (2008a) Fast response air-to-fuel ratio measurements using a novel device based on a wide band lambda sensor. Meas Sci Technol 19(075201)

Regitz S, Collings N (2008b) Study of cycle-by-cycle air-to-fuel ratio determined from the exhaust gas composition and a novel fast response device based on a wide band lambda sensor. In: SAE technical paper 2008–01-2439

Riegel J, Neumann H, Wiedenmann HM (2002) Exhaust gas sensors for automotive emission control. Solid State Ion 152–153:783–800

Roberts C (2011) The pursuit of high efficiency engines-SwRI programs. In: Emissions conference (2011), Ann Arbor, MI

Saidur R, Rezaei M, Muzammil WK, Hassan MH, Paria S, Hasanuzzaman M (2012) Technologies to recover exhaust heat from internal combustion engines. Renew Sustain Energy Rev 16(8):5649–5659

Schilling A (2008) Model-based detection and isolation of faults in the air and fuel paths of common-rail DI diesel engines equipped with a lambda and a nitrogen oxides sensor. PhD Thesis, ETH-Zürich

Schilling A, Amstutz A, Onder CH, Guzzella L (2006) A real-time model for the prediction of the NO_x emissions in DI diesel engines. In: Proceedings of the 2006 IEEE international conference on control applications, Munich, Germany

Schilling A, Amstutz A, Guzzella L (2008) Model-based detection and isolation of faults due to ageing in the air and fuel paths of common-rail direct injection diesel engines equipped with a λ and a nitrogen oxides sensor. In: Proceedings of the Institution of Mechanical Engineers. Part D: J Automobile Eng 222:101–117

Schommers J, Duvinage F, Stotz M, Peters A, Ellwanger S, Koyanagi K, Gildein H (2000) Potential of common Rail injection system for passenger car DI diesel engines. In: SAE technical paper 2000–01-0944

Schönauer D, Nieder T, Wiesner K, Fleischer M, Moos R (2011) Investigation of the electrode effects in mixed potential type ammonia exhaust gas sensors. Solid State Ion 192(1):38–41

Shutty J, Benali H, Daeubler L, Traver M (2007) Air system control for advanced diesel engines. In: SAE technical paper 2007–04-16

Simon D (2006) Optimal state estimation: Kalman, H infinity, and nonlinear approaches. Wiley, New York

Smith JA (2000) Demonstration of a fast response on-board NO_x sensor for heavy-duty diesel vehicles. SwRI Project No. 03–02256 Contract No. 98–302. In: Technical report, Southwest Research Institute Engine and Vehicle Research Division PO Box 28510 San Antonio, Texas 78228–0510

Stefanopoulou AG, Kolmanovsky I, Freudenberg JS (2000) Control of variable geometry turbocharged diesel engines for reduced emissions. IEEE Trans Control Syst Technol 8(4):733–745

Steppan J, Henderson B, Johnson K, Yusuf Khan M, Diller T, Hall M, Lourdhusamy A, Allmendinger K, Matthews R (2011) Comparison of an on-board, real-time electronic PM sensor with laboratory instruments using a 2009 heavy-duty diesel vehicle. In: SAE technical paper 2011–01-0627

Stewart G, Borrelli F, Pekar J, Germann D, Pachner D, Kihas D (2010) Toward a systematic design for turbocharged engine control, vol 402. Lecture notes in control and information sciences. Springer, London

Stiller C, Puente León F (2011) Information fusion for automotive applications—An overview. Inf Fusion 12(4):244–252

Stotsky A, Kolmanovsky I (2002) Application of input estimation techniques to charge estimation and control in automotive engines. Control Eng Pract 10(12):1371–1383

Surenahalli HS, Parker GG, Johnson JH, Devarakonda MN (2012) A Kalman filter estimator for a diesel oxidation catalyst during active regeneration of a CPF. In: Proceedings of the American control conference, pp 4969–4974

Takagi T, Sugeno M (1985), Fuzzy identification of systems and its applications to modeling and control. IEEE Trans Syst Man Cybern SMC-15, no 1

Timoney DJ, Desantes JM, Hernández L, Lyons CM (2005) The development of a semi-empirical model for rapid NO_x concentration evaluation using measured in-cylinder pressure in diesel engines. In: Proceedings of the Institution of Mechanical Engineers. Part D: J Automobile Eng 219(5):621–631

Tobias P, Msrtensson P, Göras A, Lundström I (1999) Moving gas outlets for the evaluation of fast gas sensors. Sens Actuators B: Chem 58(1–3):389–393

Trimboli S, Di Cairano S, Bemporad A, Kolmanovsky IV (2012) Model predictive control with delay compensation for air-to-fuel ratio control. Lecture notes in control and information sciences, vol 423. Springer, Berlin

Tschanz F, Amstutz A, Onder CH, Guzzella L (2012) Feedback control of particulate matter and nitrogen oxide emissions in diesel engines. Control Eng Pract (in press)

TuneståI P, Hedrick JK (2003) Cylinder air/fuel ratio estimation using net heat release data. Control Eng Pract 11(3):311–318

Turner JD (2009) Automotive sensors. In: Sensor technology series. Momentum

Twigg MV (2007) Progress and future challenges in controlling automotive exhaust gas emissions. Appl Catal B: Environ 70(1–4):2–15

van Basshuysen R, Schaefer F (2004) Internal combustion engine handbook—Basics. SAE Int Comp Syst Perspect

van Nieuwstadt M (2003) Coordinated control of EGR valve and intake throttle for better fuel economy in diesel engines. In: SAE technical paper 2003–01-0362

van Nieuwstadt M, Upadhyay D (2005) Diagnosis of a urea SCR catalytic system

Varnier O (2012) Trends and limits of two-stage boosting systems for automotive diesel engines. PhD Thesis, Universitat Politècnica de València, Departamento de Máquinas y Motores Térmicos

Viskup R, Alberer D, Oppenauer K, del Re L (2011) Measurement of transient PM emissions in diesel engine. In: SAE technical paper 2011–24-0197

Vogt M, Müller N, Isermann R (2004) On-line adaptation of grid-based look-up tables using a fast linear regression technique. J Dyn Syst Meas Control, Trans ASME 126(4):732–739

Wagner JF, Wieneke T (2003) Integrating satellite and inertial navigation—Conventional and new fusion approaches. Control Eng Pract 11(5):543–550 (Automatic Control in Aerospace)

Wahlström J (2009) Control of EGR and VGT for emission control and pumping work minimization in diesel engines. PhD Thesis, Linköpings Universitet. LiU-TEK-LIC-2006:52, Thesis no 1271

Wang DY, Yao S, Shost M, Yoo J-H, Cabush D, Racine D, Cloudt R, Willems F (2008) Ammonia sensor for closed-loop SCR control. In: SAE technical paper 2008-01-0919

Wang DY (2007) Real-time dynamics of amperometric exhaust oxygen sensors. Sens Actuators B: Chem 126(2):551–556

Warey A, Hall MJ (2005) Performance characteristics of a new on-board engine exhaust particulate matter sensor. In: SAE technical paper 2005–01-3792

Westlund A (2011) Simplified models for emission formation in diesel engines during transient operation. PhD Thesis, KTH Industrial Engineering and Management

Westlund A, Winkler N, Diotallevi F, Ångström H (2008) Predictions and measurements of transient NO emissions for a two-stage turbocharged HD diesel engine with EGR. In: Thiesel conference on thermo-fluid dynamics processes in diesel engines, Valencia, Spain

Winkler-Ebner B, Hirsch M, del Re L, Klinger H, Mistelberger W (2010) Comparison of virtual and physical NO_X-sensors for heavy duty diesel engine application. SAE Int J Engines 3(1): 1124–1139

Woschni G (1967) A universally applicable equation for the instantaneous heat transfer coefficient in the internal combustion engine. In: SAE technical paper 670931

Wu G (2006) A table update method for adaptive knock control. In: SAE technical paper 2006-01-0607

Yan F, Wang J (2012) Pressure-based transient intake manifold temperature reconstruction in diesel engines. Control Eng Pract 20(5):531–538

Yang H, Tsourapas V, Prakash AK, Yuan Q, Rugge R, Mhaskar U, Rao PN (2009) Hardware-in-the-loop (HIL) modeling and simulation for diesel aftertreatment controls development. In: SAE technical paper 2009–01-2928

Yen G, Michel AN (1991) A learning and forgetting algorithm in associative memories: results involving pseudo inverses. In: IEEE international symposium on circuits and systems, vol 2, pp 778–781

Zeldovich J (1946) The oxidation of nitrogen combustion and explosions. Acta Physicochim 21(4):577–628

Zhou G, Jørgensen JB, Duwig C, Huusom JK (2012) State estimation in the automotive SCR deNO_X process. In: IFAC proceedings volumes, vol 8, pp 501–506

Zhuiykov S (2010) Electrochemistry of zirconia gas sensors. Taylor & Francis, London

Zhuiykov S, Miura N (2007) Development of zirconia-based potentiometric NO_X sensors for automotive and energy industries in the early 21st century: what are the prospects for sensors? Sens Actuators B: Chem 121(2):639–651

Part II
Sensors and Models for NO_x and λ^{-1}

Chapter 3
System Setup and Sensors Characterisation

3.1 Introduction

This chapter is divided into two parts. The first part, which comprises Sects. 3.2 and 3.3, is devoted to the description of the experimental configuration and tests used for this work. Section 3.2 describes the experimental set-up used in the present work, including the relevant characteristics of the engine, sensors and test cell equipment, while the Sect. 3.3 presents the steady-state and dynamic tests performed to tune and validate the methods.

In the second part of the chapter, the use of online methods for characterising NO_x and λ^{-1} output from exhaust gas concentration sensors is emphasised. NO_x output is characterised by a novel method based on SOI steps (Galindo et al. 2011), while λ^{-1} output is characterised by performing injection steps (Guardiola et al. 2013).

3.2 Experimental Set-up

Test engine is an automotive 2.2-litre 4-cylinder common rail DICI engine with a sequential parallel turbocharger Galindo et al. (2007), certified according to EURO 4 standard (model DW12B from PSA). Engine specifications are shown in Table 3.1. A bypass-allowed ECU permits varying injection parameters, such as p_{rail}, SOI and main injection duration (t_{mi}), and also boost and EGR control set-points. The engine is installed on an engine test bench and coupled to a variable frequency eddy current dynamometer that allows carrying out dynamical tests. Standard air intercooler is substituted by a water intercooler for a proper heat extraction on the intake air mass flow. The usual experimental set-up used for generating the data in this work is shown in Fig. 3.1, where the AT devices are removed and the main sensors are sketched. A backpressure valve is installed downstream of the turbine outlet and commanded for replicating the pressure drop in the exhaust line of series engines.

D. Blanco-Rodriguez, *Modelling and Observation of Exhaust Gas Concentrations for Diesel Engine Control*, Springer Theses, DOI: 10.1007/978-3-319-06737-7_3,

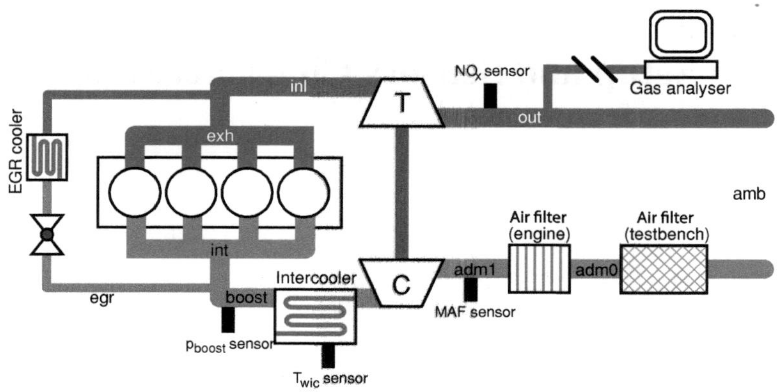

Fig. 3.1 Engine layout with the main sensors used in this work (pressures, temperatures in different positions are also measured but are not shown for clarity)

Table 3.1 Main characteristics of the DW12B engine from PSA

Stroke (S)	96 mm
Bore (D)	85 mm
S/D	1.129
Number of cylinders	4
Displacement	2179 cm^3
Turbocharging system	Sequential parallel
EGR	HP
Injectors	Solenoid (common-rail)
Valves by cylinder	4
Maximum power	125 kW @ 4,000 rpm
Compression ratio	17:1

Table 3.2 Main actuators used in the experimental setup

EGR command	u_{egr}	%
WG command	u_{wg}	%
Rail pressure command	$u_{p_{rail}}$	bar
Main injection duration (MI)	$u_{t_{mi}}$	μs
Start of injection of the MI	u_{soi}	°BTDC

The nomenclature for the engine positions is clarified in the Fig. 3.1 and in the list of variables at the beginning of the document as well. In some specific tests, the sensor layout is modified and AT devices are included, however they are indicated where proceed. The main actuators on the engine are described in the Table 3.2.

The Rapid Prototyping System

In order to avoid the shifting and causality problems with the signals, the system layout shown in the Fig. 3.2 is used for acquiring from and commanding signals to the engine. A real time (RT) PXI system (from National Instruments) is used for

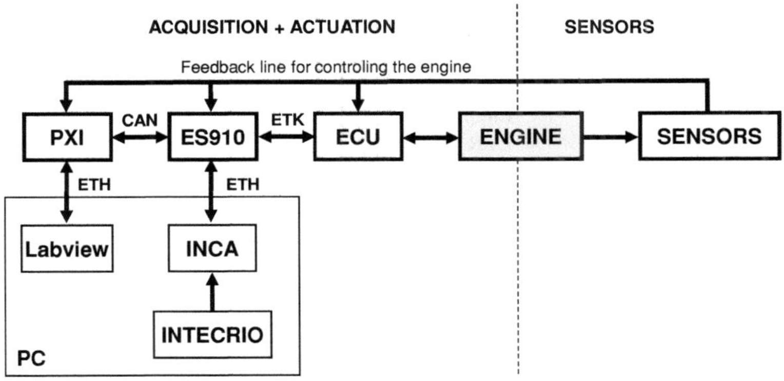

Fig. 3.2 System layout for rapid prototyping (fast acquisition, bypass and control) and integration of new sensors for testing the engine

the acquisition of data from series and experimental sensors. The PXI is connected via CAN with a ES910 hardware from ETAS for acquisition and bypass of the ECU signals, allowing to design control routines and skipping ECU logic. In addition to series sensors (already connected with the ECU), other sensors are installed on engine and linked with the PXI for acquisition and providing feedback for control strategies. Labview software is used for managing PXI, while INCA software is utilised for interfacing with the ES910. INTECRIO software permits to program the engine variables for the bypass.

When bypassing, the time needed by the PXI software for writing into the ECU channel is 20 ms, and the same for reading in the other direction. Then, the maximum expected error between commanding and reading a signal in the PXI is 40 ms for the round path. Lags are around 20 ms when storing data by INCA software (used for sensors characterisation and connected with the ES910). Acquisition frequencies are 100 Hz for integrated sensors and between 10 and 100 Hz for ECU signals (fixed by manufacturer). Commanding frequencies depend on the test and variable.

3.2.1 Engine Sensors

Besides the commercial sensors, a full set of sensors is installed on-engine in order to measure full operating conditions (pressures, temperatures, mass flows, engine and turbo speeds, torque and fuel consumption), atmospheric conditions and emissions. Some of them are measured redundantly for calibration. All sensors are tested and calibrated with standard procedures, both in steady and transient-state. Table 3.3 shows the main series and test cell sensors harnessed for measuring engine conditions. Due to the relevance in this work, gas concentration sensors are described separately.

Table 3.3 Sensors used in the experimental setup (gas concentration sensors are discussed bellow)

Series sensors	Symbol	Location
Air mass flow (MAF) (kg/h)	$\dot{m}_a$	*adm1*
Boost pressure (bar)	p_{boost}	*boost*
Engine speed (rpm)	n	*crank*
EGR valve position (%)	x_{egr}	*EGR valve*
WG position (%)	x_{wg}	*WG*
Rail pressure (bar)	p_{rail}	*CR*
Test cell sensors		
Torque (Nm)	M_e	*crank*
Air mass flow (sensyflow) (kg/h)	$\dot{m}_a^{cell}$	*adm0*
Fuel (gravimetric) (kg/h)	$\dot{m}_f^{cell}$	*external*
Pressures (bar)	p	*runners*
Temperatures (°C)	T	*runners*
In-cylinder pressure (bar)	p_{cyl}	*cyl*

3.2.1.1 Exhaust Gas Concentrations Sensors

Different gas concentration sensors are installed for designing and validating the algorithms. These can be summed up in research grade and on-board (experimental and commercial) sensors. From the first type, a gas analyser is used for measuring gas concentrations and an AVL 439 opacimeter for measuring the opacity at the exhaust (not relevant in this work). From the second type, λ sensors and NO_x sensors are highlighted. When necessary, redundancy is applied to sensors for comparing signals in different locations and indeed different units, e.g. the measurement of λ^{-1} before and after the turbine. Appendix A.1 summarises the technical data of these sensors according to the suppliers.

Gas Analyser

A Horiba MEXA 7100DEGR gas analyser HORIBA (2001) is used as calibration standard and among others, the following relevant concentrations are measured: CO_2 at the intake and exhaust, NO_x at the turbine outlet and O_2 at the intake manifold and turbine outlet. The engine-out λ^{-1} and NO_x are inferred by locating the gas probe downstream of the turbine. λ^{-1} is calculated from the O_2 measurement, estimated by means of a magneto-pneumatic detection (MPD), while NO_x is measured on a dry basis, by means of a heated chemiluminescence detector (HCLD) with a NO2/NO converter. Calibration gases are used for the calibration of the system (zero and span).

Three key factors do not allow to use this system as a dynamic calibration method for on-board sensors:

1. Measurement system response time, which includes the signal rising time, is slower than the tested on-board sensors.
2. A long line is used for providing exhaust gas to the gas analyser (12 m). This long line causes not only a transport delay, but also can cause diffusive effects.
3. Soot-free gas is needed in order to prevent sensor damage; therefore a filter is used in the line Bermúdez et al. (2008). In transient engine operation, filter contributes to substantial emission signal distortion.

It is worth noting that some studies combine experimental data and phenomenological modelling approach to synchronise Broatch et al. (2008); Arrègle et al. (2006) or to restore Chan et al. (1997); Geivanidis and Samaras (2008); Mrosek et al. (2011) the signals from slow response analysers during transient engine operation. This is done by analytical means, using simple gas diffusion models and perfect gas mixing models, or through a control theory approach to the problem. Section 3.4 shows static and dynamic results by using the gas analyser for characterising the NO_x sensor.

UEGO Sensor for Measuringλ^{-1}
This on-board sensor may be located in different positions in order to evaluate the possibilities of using measured λ^{-1} in models and adaptive algorithms. The proposal in this work for estimating engine-out λ^{-1} is utilising a NO_x sensor, capable of measuring also λ^{-1}, while avoiding the use of a UEGO sensor (used for some production engines). Anyway, the sensor is used for helping some explanations and as a reference for comparison with the NO_x sensor. Section 3.2.1 presents different alternatives for measuring intake λ^{-1} from UEGO sensors and the gas analyser.

ZrO_2*-Based Sensor for Measuring* λ^{-1} *and* NO_x
The availability and the trade-off for locating NO_x sensors in production diesel engine cars varies depending on the final application. In HD engines, the NO_x sensor is becoming usual, and the normal solution is locating one NO_x sensor downstream of the SCR for diagnosing and managing the system, while exhaust NO_x is observed (saving the need of using more than one sensor unit). In LD, the sensor will become necessary for EURO 6 diesel engines with a similar trade-off to that of HD (although LNT might be used instead of SCR), while some luxury cars already mount two NO_x sensors for providing a valid reference before and after the AT devices.

In this work, the AT devices are removed in the standard configuration and the sensor is located at the turbine outlet. The capability of measuring λ^{-1} permits to work without UEGO sensors. It should be clarified that even though λ^{-1} is not really necessary for the control of diesel engines, it is advised for emissions control (soot and NO_x) and may be used for correcting injector drift or bounding injection (on the basis of the smoke limiter actuation). Anyway, the sensor response must be assessed and this is performed in Sect. 3.4.

3.2.1.2 Intake Gas Concentration Sensors

Due to the strong dependency of NO_x emissions on $[O_2]_{int}$, the measurement of $[O_2]_{int}$ is of major importance. However, the lack of homogeneity of the intake mix stands a major problem in some engines. In this work, the methodology presented in Luján et al. (2008) has been used for analysing the homogeneity of the mix. The method is based on feeding the gas analyser from different probes, as sketched in Fig. 3.3:

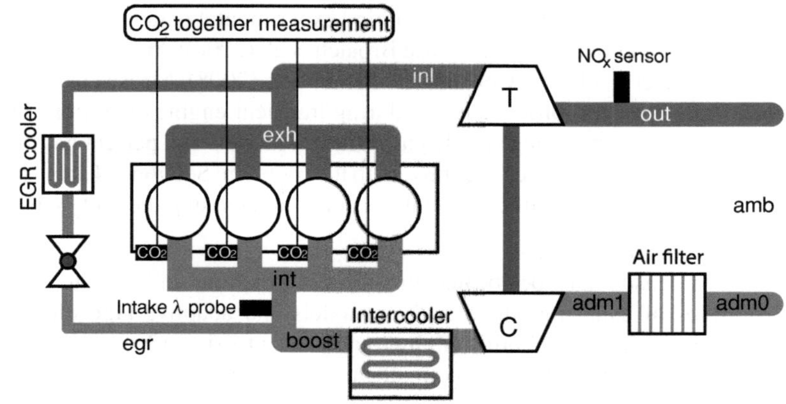

Fig. 3.3 Engine layout showing the main sensors used for measuring $[O_2]_{int}$ or equivalently $[CO_2]_{int}$

- Measurement of the $[CO_2]_{int}$ just before the separation of the manifolds at the intake by the gas analyser,
- extractions in the four individual cylinder intake runners and measurement by the gas analyser (cylinder 1 to 4 $[CO_2]_{int}$ independent measurements),
- measurement of a sample result of the mix of the 4 extractions and made by the gas analyser (1 measurement, from a rail where gas coming from all the probes is mixed).

When using the gas analyser, the $[CO_2]_{int}$ (%) is measured with a non-dispersive infrared method (NDIR). As the intake gas is a mix of ambient air and the results of the combustion of this air with fuel, the $[O_2]_{int}$ is linearly correlated with the $[CO_2]_{int}$ (the higher the $[CO_2]_{int}$, the lower the $[O_2]_{int}$).

Additionally, and as a real time alternative, the measurement of the intake λ by means of the UEGO sensor located in the intake manifold has been implemented. The sensor has been located with a certain distance upstream the separation between individual cylinder runners for avoiding separation effects (1 measurement).

UEGO sensor provides the ion pump current I_p (mA). The I_p signal is previously corrected with the effect of p_{boost} which affects to the oxygen partial pressure measurement. Furthermore, I_p is related with λ^{-1} (and thus $[O_2]_{int}$ and $[CO_2]_{int}$) by a curve given by the manufacturer.

A set of measurements is made in order to evaluate the different alternatives and to determine the best method to estimate the intake gas concentration. In total 4 operating points defined by the pairs $[n, m_f]$ are studied: [2,500, 35], [2,500, 7], [1,500, 20] and [1,500, 7] (units are rpm and mg/str respectively) with the rest of inputs fixed by the standard engine calibrations.

The results of the study are summarised in Table 3.4 for the 4 operating points. Even though there exists an important dispersion among the cylinder samples (due to not utilizing a mixer that guarantees the homogeneity of the fluid at intake), the

Table 3.4 Comparison of different sensor outputs for determining the effective intake $[CO_2]_{int}$ (and thus $[O_2]_{int}$ and λ^{-1}) in the cylinder

Operating point $[n, m_f]$	[2,500, 35]	[2,500, 7]	[1,500, 20]	[1,500, 7]
$[CO_2]_{int}$ (%)	1.37	1.07	1.27	2.30
$[CO_2]_{int}$ 1 (%)	1.14	0.98	2.38	2.62
$[CO_2]_{int}$ 2 (%)	1.65	1.09	2.46	2.92
$[CO_2]_{int}$ 3 (%)	2.07	1.11	1.89	2.64
$[CO_2]_{int}$ 4 (%)	1.89	1.10	1.79	2.55
$[CO_2]_{int}$ 1–4 (%)	1.62	1.07	2.27	2.73
$[CO_2]_{int}$ mean 1–4 (%)	1.69	1.07	2.13	2.68
std 1–4 (%)	0.40	0.06	0.34	0.16
I_p (mA)	1.93	2.12	1.91	1.78
$[CO_2]_{int}$ (%) from I_p	1.99	0.95	2.10	2.80

std standard deviation

The first row represents the measurement of $[CO_2]_{int}$ by a gas analyser just upstream of the manifolds at the intake junction; the rows 2–5 represent the gas analyser measurements of $[CO_2]_{int}$ by means of the 4 individual extractions of gas at the intake manifolds of the engine; the row 6 is the gas analyser measurement of the sample result of a mix of 4 equal extractions of the intake manifolds of each cylinder; rows 7–8 represents the mean and standard deviation of the 4 individual extractions of the gas measured in rows 2–5; while the last two rows represents the measurement of the intake lambda sensor.

Fig. 3.4 Equivalency between $[CO_2]_{int}$ and the pressure corrected signal I_p from the UEGO sensor. The saturation of the intake probe is evident for values of I_p around 2.3, i.e. equivalent to a low $[CO_2]_{int}$ concentration

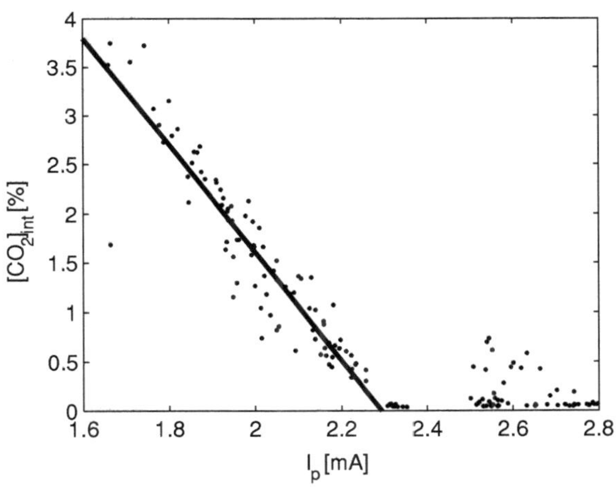

comparison of the mean value for the 4 cylinders with the measurement from the rail (Cyl 1-4 probe) is satisfactory. Nevertheless, the $[CO_2]_{int}$ measured just before the cylinder manifolds is highly affected by the lack of homogeneity in the mix and the signal presents a significant error when comparing with the rest of alternatives. The results obtained by the UEGO sensor are acceptable, despite of a slight deviation lower than that of the gas analyser intake extraction. However, the Bosch sensor presents a saturation when $I_p > 2.3$ mA (low $[CO_2]_{int}$ and high $[O_2]_{int}$) as shown in Fig. 3.4. Chapter 4 completes the discussion comparing the different possibilities (sensors and models) for estimating the $[O_2]_{int}$.

3.3 Engine Tests

Steady-state and dynamic tests are performed for obtaining the results presented in this work. A summary on these is shown in this section.

3.3.1 *Steady-State Tests*

A full set of steady-state tests are performed in order to characterise the engine and fit the tables and parameters of the models described in Chap. 4. This includes different pairs of $[n, m_f]$ values covering:

- nominal tests, defined by the manufacturer ECU settings, with warm conditions and variables stabilised,
- variations of the EGR rate (EGR),
- variations of the engine coolant temperature (T_{cool}),
- variations of the water temperature of the intercooler (T_{wic}) to influence the intake temperature (T_{int}), and
- variations of p_{boost} to influence on $\dot{m}_{int}$.

Ranges are selected in order to produce appreciable NO_x variations but taking into account the limits for engine operation. 363 steady measurements are made in total. Covered engine areas are shown in Fig. 3.5.

3.3.2 *Transient Tests*

3.3.2.1 Steps

Step-like tests are used for identifying sensor and system dynamics (flow, temperatures and pressures). Table 3.5 shows specific tests performed for different engine operating points.

Furthermore and for demonstrating the effect of sharp variations in SOI on the NO_x emissions, 19 specific tests consisting in several consecutive steps were performed at different engine speeds (from 1,000 to 3,000 rpm) with different initial NO_x concentrations (i.e. different initial engine operating conditions) and step sizes (ranging from 1° to 4° in SOI). Regarding the fuel-to-air ratio, specific load transients are performed by keeping the EGR valve fully closed in order to avoid possible interactions. Data is later used for the sensor dynamics identification in Sect. 3.4.

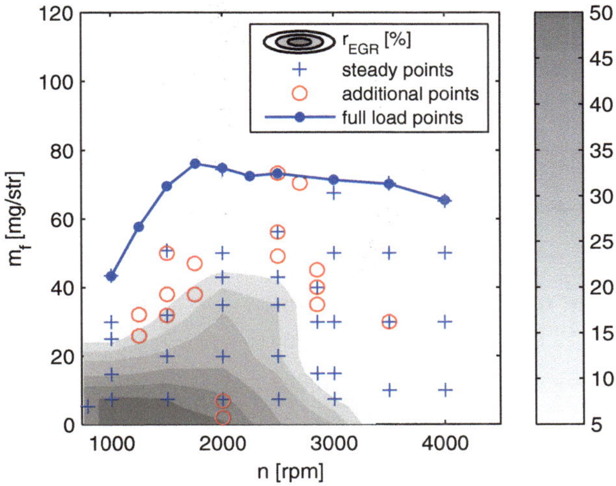

Fig. 3.5 $[n,\ m_f]$ set of values run for testing the engine in steady-state. In each set, different variations of u_{egr}, T_{cool}, p_{boost} and T_{int} are made for characterizing the engine. Some additional points are measured for completing the engine characterization. Nominal EGR map EGR_0 (white area corresponds to EGR valve closed, $EGR = 0$) and full load line are also plotted

Table 3.5 Steps performed in the engine for system and sensors characterisation

n (rpm)	m_f (mg/str)	Δu_{soi} (° BTDC)	Δu_{egr} (%)	Δu_{wg} (%)
1,000	7.5	±2	± 10	−30
1,000	43	±2		−3
2,500	35	±2	±5	±3
4,000	10	±2		±5
4,000	50	±2		±3

BTDC before top dead centre

3.3.2.2 Cycles

Different engine cycles are designed on the basis of the homologation cycle and *ad-hoc* modifications for proving and testing the models and methods developed in this work. In the following, the cycles are described and the test campaign is summed up in Table 3.7.

Adapted Transient Cycles (TRAN)
A set of dynamic tests are designed on the basis of two different modifications of the FTP cycle for HD engines (named here as TRAN A and TRAN B), and are used for training and evaluating a NO_x model in transient operation. These consist of a re-escalation of the central part of the standard FTP and are compared in Fig. 3.6. Test proposition is a full factorial design by performing modifications on set-points for u_{egr} (by modification on the $\dot{m}_a$ maps), p_{boost} (by fixing $\dot{m}_a$ set-point and modifying p_{boost} map) and T_{cool} (modifying the engine coolant reference temperature at

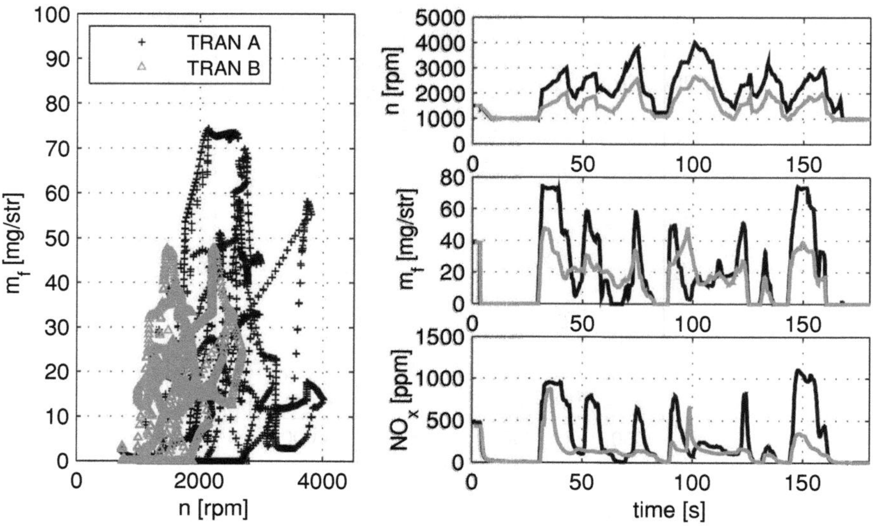

Fig. 3.6 TRAN nominal cycles designed for the model validation. TRAN A is the black line, while TRAN B is the grey line

Table 3.6 Experimental plan of the modified TRAN cycles

Cycle	T_{cool} (°C)	Δu_{egr}(%)	Δp_{boost} (mbar)
TRAN A	[75, 85, 95]	[−5, 0, +5]	[−100, 0, +100]
TRAN B	nom	[−10, 0, +20]	[−100, 0, +100]

the test cell controller). The sequence is presented in Table 3.6, totalling 36 different tests (27 TRAN A and 9 TRAN B, both in the test campaign A). For each test, three repetitions of the sequence shown in Fig. 3.6 are made.

The NO_x sensitivity to variations on inputs and parameters is considerably lower for transient than static conditions. This fact is due to the systems and sensors dynamics which filter the effects. For instance, an increasing of 10 % in p_{boost} (100 mbar) in the TRAN A cycle leads to an increase NO_x around 10 %, while in steady conditions produces an increase in the order of 40 % for the same conditions (while fixing $\dot{m}_a$ set-point in the EGR area of the engine). In all cases, the higher NO_x sensitivity is found for EGR variations, especially for TRAN B cycles, which present sharper EGR transients. Figure 3.7 shows the effect of modifying EGR rate (by plotting $[CO_2]_{int}$) over the final NO_x emissions measured by the sensor.

Sportive Driving Mountain Profile (SDMP)

The SDMP is designed for covering a wide operating range of the engine by sharp variations on n and m_f. A zoom on 400 s (total is 1,200) of the SDMP is shown in Fig. 3.8 (left plot indicates the whole cycle points). Since the SDMP presents

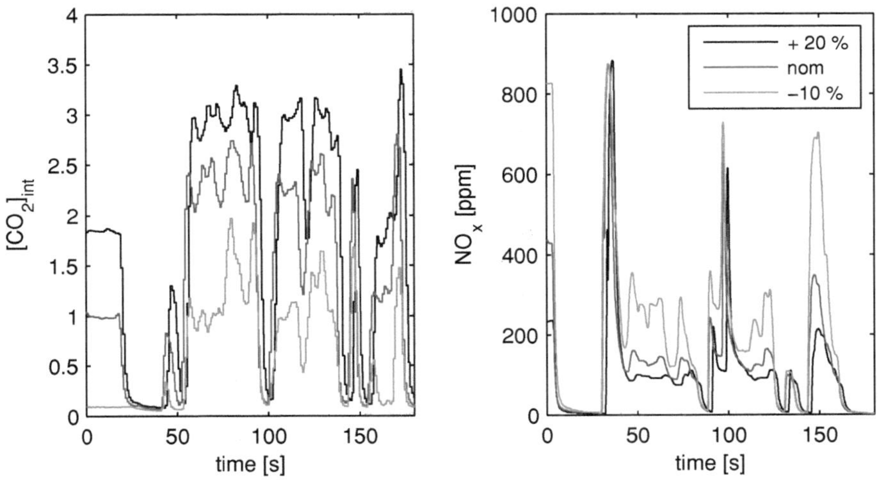

Fig. 3.7 $[CO_2]_{int}$ at intake and NO_x emissions for TRAN B after varying the EGR set-point. The legend in the right plot shows the values of u_{egr}. Note that the higher u_{egr} the lower $\dot{m}_{egr}$

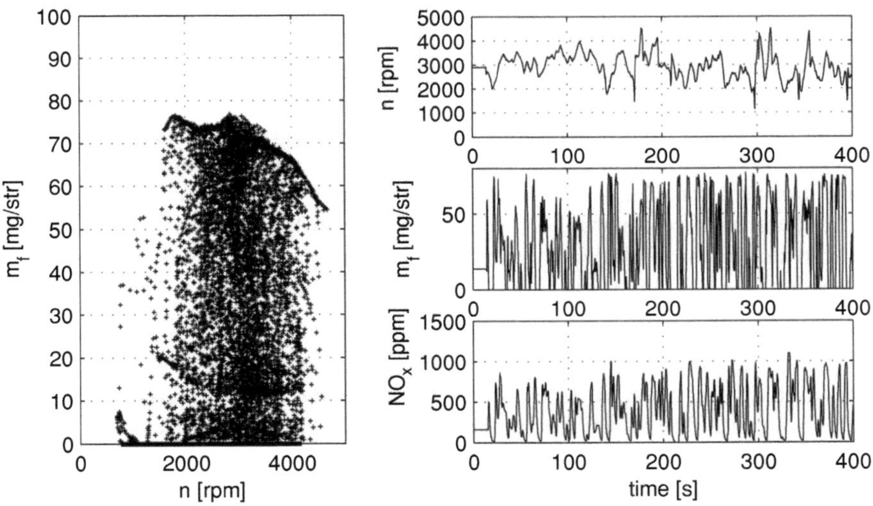

Fig. 3.8 SDMP cycle performed on the engine

sharp load transients, the EGR valve is mostly closed during the cycle, making that $EGR = 0$ during the major part in spite of the engine might be run in the EGR area. This circumstance distorts the closed loop nominal engine operation.

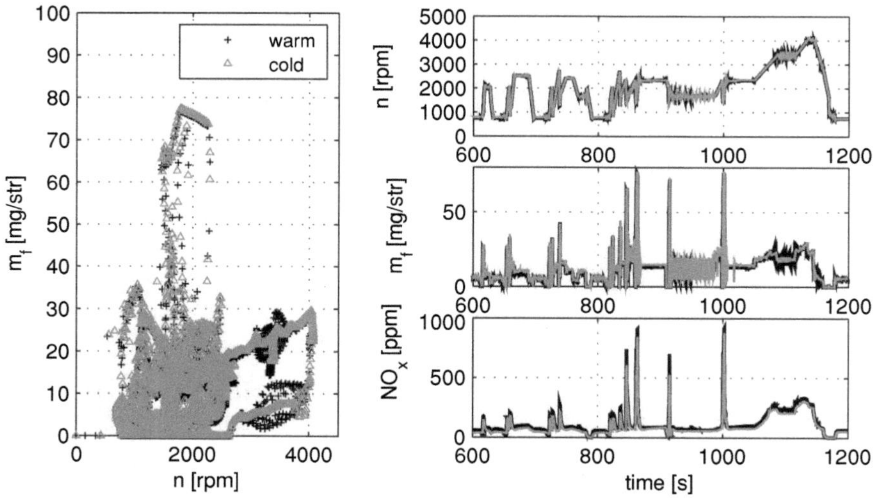

Fig. 3.9 NEDC cycle performed on the engine

New European Driving Cycle (NEDC)
The NEDC is tested in cold and warm-start conditions (at the beginning of the test). The cycle consists of a urban part with accelerations/decelerations and idling, and an extra-urban part that lasts 400 s. The urban part has 4 equal repetitions of a sequence that lasts 200 s; Fig. 3.9 shows last repetition of the urban part and the extra-urban section. Cold-start cycle presents lower NO_x, indeed when m_f is slightly higher because of cold starting strategies. Anyway, differences are not considerable and can also be associated with cycle differences in operating points (n and m_f have dispersion, see left plot of Fig. 3.9).

Common Artemis Driving Cycles (CADC)
CADC cycles are designed upon the basis of an statistical analysis of European real world driving patterns, developed by the European Artemis[1] project. The cycle includes three different variants: urban, rural road and highway. These profiles are more realistic than the NEDC as they cover different real situations and then constitute a good source for experimental testing. Three versions of the CADC are shown in Figs. 3.10–3.12.

Summary on the Transient Cycles
In total, two different test campaigns for testing dynamic cycles have been performed for the present work; these are separated in test campaign A and B just for remarking that they have been performed in different moments. Table 3.7 sums up the dynamic tests made, where an identifier (ID) names the cycles. A3 to A29, and A30 to A38 stand for the test of the full factorial design summarised in Table 3.6.

[1] ARTEMIS: Assessment and Reliability of Transport Emission Models and Inventory Systems.

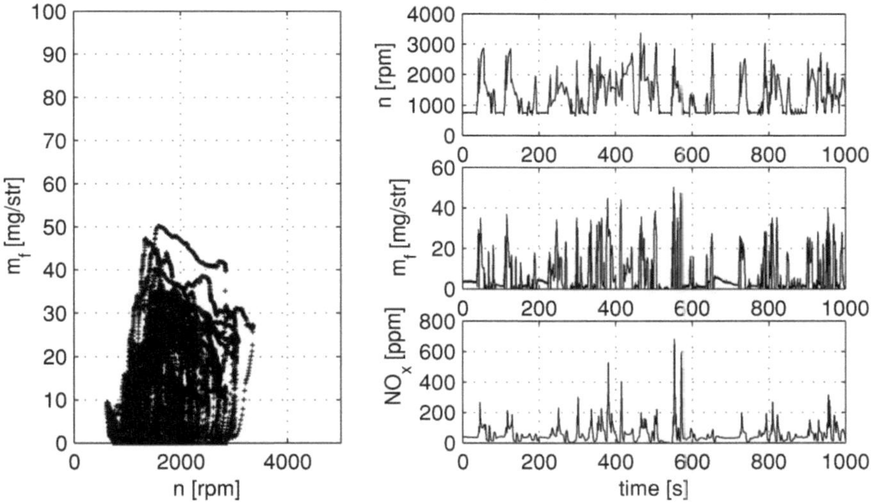

Fig. 3.10 Urban CADC cycle performed on the engine.

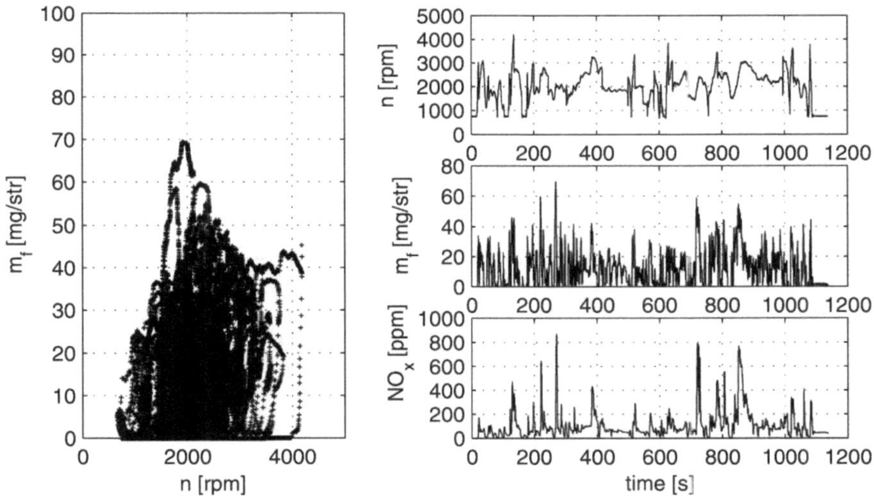

Fig. 3.11 Rural CADC cycle performed on the engine

3.4 Gas Concentration Sensors Characterisation

In the next, the static and dynamic calibration of λ and NO_x output from the gas concentration sensors is discussed. The static calibration is based on the linear fitting of sensor outputs by using the steady-state data-set collected from the engine, while

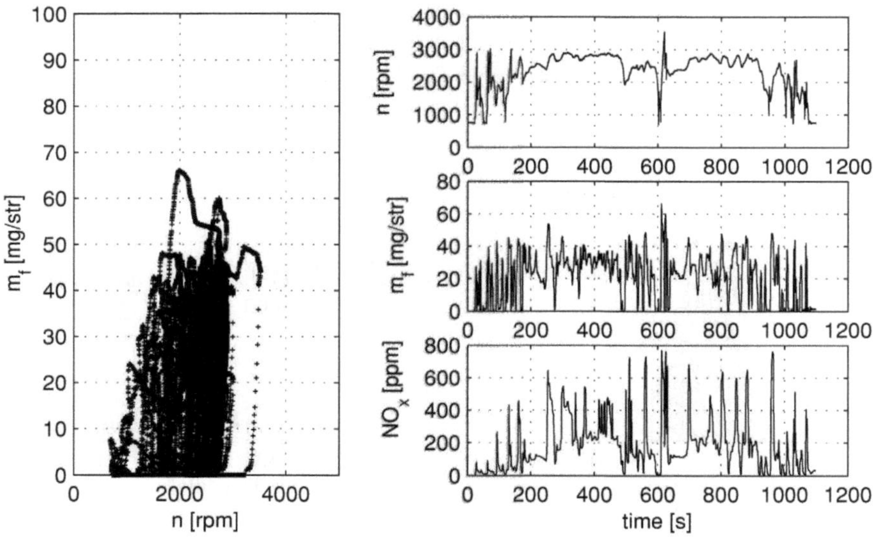

Fig. 3.12 Highway CADC cycle performed on the engine

Table 3.7 Dynamic test campaigns

ID	Cycle	Calibration
Test campaign A		
A1	NEDC warm	nom
A2	NEDC cold	nom
A3-A29	TRAN A	DoE
A30-A38	TRAN B	DoE
Test campaign B		
B1	NEDC warm	nom
B2	CADC Urban	nom
B3	CADC Rural	nom
B4	CADC Highway	nom
B5	SDMP 1	nom
B6	SDMP 2	nom

the dynamic characterisation is based on generating exhaust gases with different concentrations by means of the step-like variations on the engine injection settings.

With respect to the NO_x output, this section presents an easy on-engine experiment for NO_x sensors by abrupt changes on the start of injection (SOI). The results shown in this section corresponds to a NO_x sensor and a gas probe located downstream of the turbine, removing the AT line.

Regarding the λ^{-1} output characterisation, variations on the injected fuel mass rate $\dot{m}_f$ are used for provoking step-like transitions on the λ output and the methodology for identifying the sensor is similar to that of the NO_x output. Figure 2.5 already compared λ^{-1} output from the NO_x and a UEGO sensor in a configuration different from the standard of this work, where the former is located downstream of the

ATs while the latter at the turbine outlet. The NO_x sensor provides a slower and more filtered response than the other due to the position and the sensor response time. However, if using a NO_x sensor, the UEGO sensor can be removed instead of using two different sensor units. In the following, the results for the NO_x sensor output located at the turbine outlet are presented.[2] The conclusions and results of the characterisation are required for the implementation of models and observers in subsequent chapters.

3.4.1 Static Calibration

Previous to the NO_x sensor static calibration, some issues must be considered. On one hand, the measurement procedures of the sensors affect the output itself, especially for NO_x where a number of different species (NO and NO_2 among others) can be distinguished. This discussion is moved to the Appendix A.2.

On the other hand, the NO_x sensor principle is based on the measurement of the partial pressure of oxygen in order to determine both λ^{-1} ($z_{\lambda^{-1}}$) in a first chamber (outer cavity) and NO_x (z_{NO_x}) in a second one (inner cavity), as shown in Fig. 2.4. The total pressure of the manifolds affects the output, as occurs also on UEGO and MAF sensors, and a correction (by software) is required. A simple analysis by disposing different gas concentrations and pressures (in a static way) in a sealed chamber allows checking the pressure effect over the NO_x sensor outputs. Two gas bottles containing pure CO_2 and a calibrated mix of NO_x and N_2 are used for designing the mixes when added to atmospheric air. A ZrO_2 sensor is installed and measurements are made in steady-state.

Figure 3.13 shows iso-NO_x curves (left plot) and iso-O_2 (right plot) for different relative pressures in the chamber and for different NO_x levels (ppm) and oxygen levels (%), tested in the sealed chamber with the actual references known beforehand. The curves are obtained discharging the vessel, which ensures that the concentration is constant along each one of the tests. As it can be noticed in the plot, the oxygen output (z_{O_2}) is already pressure compensated by the manufacturer, while the NO_x output is affected and a specific correction seems necessary if the sensor is to be used with varying gas pressures.

The NO_x sensor is installed at the turbine outlet. The turbine outlet pressure (p_{out}) is in all cases between 1 and 1.6 bar (in the NEDC the maximum p_{out} is about 1.3 bar). Derived from the limited range in p_{out}, the possible error due to the pressure effect is lower than the expected error according to the specifications provided by the manufacturer (shown in Appendix A.1); hence no specific correction has been used in the present work.

[2] The UEGO sensor is used for initial phases of this work but then removed.

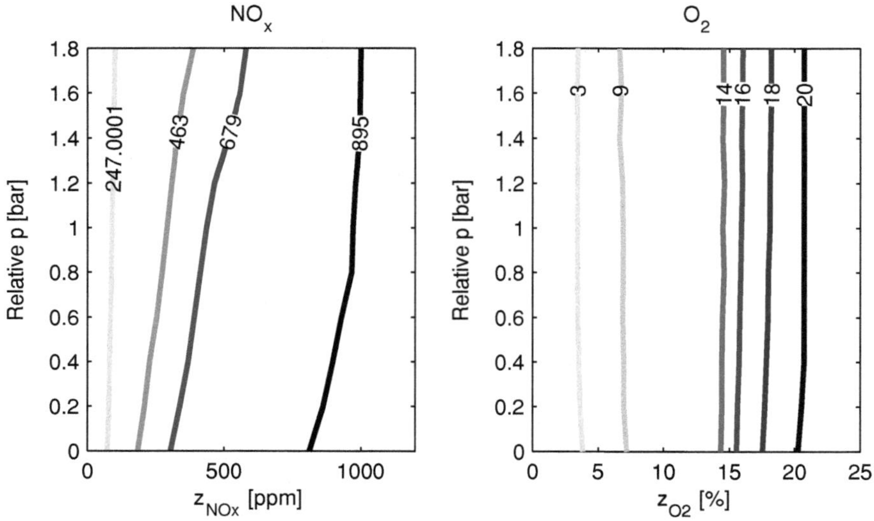

Fig. 3.13 Effect of the pressure over the NO_x sensor for the NO_x output z_{NO_x} (*left*) and oxygen estimation z_{O_2} from λ^{-1} output (*right*). It seems that the manufacturer has compensated the oxygen output but not the NO_x output

3.4.1.1 Static Calibration of the NO_x Output

With respect to the NO_x sensor accuracy, the ZrO_2 sensor is calibrated by comparing the signal with the NO_x output of the gas analyser, used as static standard. The NO_x output in the gas analyser is measured by a chemiluminescence system (CLD) in a dry basis and corrected by humidity and other factors. The important dependability of the exhaust NO_x concentration when varying the engine operating conditions is profited for the steady calibration. The steady-state tests are used for the calibration, causing NO_x concentration variations to cover the sensor measuring range. Figure 3.14 illustrates the results. After removing the outliers (usually caused by human errors in the measurement procedure of the gas analyser) and using the box-and-whiskers method, a linear fit is performed. Mean absolute error after calibrating NO_x from ZrO_2 output results of 10.7 ppm, and the residuum standard deviation of 14.77 ppm.

3.4.1.2 Static Calibration of the λ^{-1} Output

NO_x sensor provides a very accurate information of λ^{-1} in steady-state operation. Static measurements of the NO_x sensor are compared to those of the gas analyser in Fig. 3.15 by using the steady-state data-set. Mean absolute error after calibrating λ^{-1} from ZrO_2 output results of 0.0048, and the residuum standard deviation of 0.004.

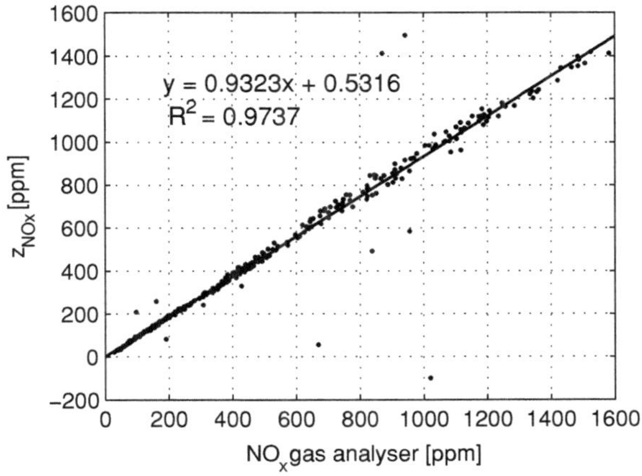

Fig. 3.14 NO_x from ZrO_2 sensor (z_{NO_x}) and the gas analyser under different steady operating conditions; the linear fit and correlation coefficient are represented

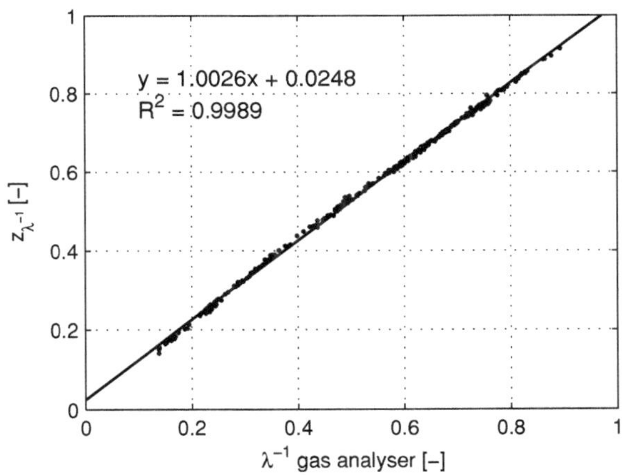

Fig. 3.15 λ^{-1} from ZrO_2 sensor ($z_{\lambda^{-1}}$) and the gas analyser under different steady operating conditions; the linear fit and correlation coefficient are represented

3.4.2 Dynamic Calibration

The second step in the calibration procedure consists in the identification of the sensor dynamics. A well-known method consists of using a step transition in the measured quantity Ogata (2001). However, usual means for providing step-like transition in gas concentration are complex set-ups with fast valves and synthetic gas mixes Regitz and Collings (2008), which can hardly being implemented on-board.

For solving this issue, on-engine alternatives are next presented for the calibration of the NO_x and λ^{-1} outputs of the exhaust concentration sensor.

3.4.2.1 Dynamic Calibration of the NO_x Output

This subsection presents a novel method based on the variation of the SOI Galindo et al. (2011), which can be easily done in current engines during their normal operation. In order to decouple as much as possible the effect of SOI variations on the air path of the engine, EGR rate is set to zero (EGR valve fully closed) during the tests. Although other variables could be used for affecting the NO_x concentration (e.g. EGR valve control), SOI is selected because:

- SOI varies the exhaust gas composition in an important way, producing appreciable NO_x concentration variations.
- SOI variations produce low variations for turbine intake temperature and pressure. Hence conditions at the turbine inlet are not importantly varied, and the turbocharged speed and boost pressure are not significantly affected; since the EGR valve is closed, the effect of the slight variation in the exhaust pressure may be neglected. This is considered as a key factor since pressure and temperature variations are associated with slow transients (because of mass and heat accumulation) which would affect the NO_x concentration.
- Other control variables correspond to slow acting valves, which will distort the step-like profile in the gas composition. However, the system response to SOI variation can be considered instantaneous: SOI is electronically applied, thus no actuation delay is expected beyond the cycle-to-cycle response time.[3] Additionally, since no EGR is performed, NO_x concentration at the cylinder exhaust port is only dependent on the trapped air mass quantity and temperature, and injection settings during the previous cycle (NO_x contained in the residual gas fraction is re-burned during the combustion Payri et al. (2008)). Finally, the gas transport delay from the cylinder to the ZrO_2 sensor (located at least 1 m from the exhaust port) is of a few tens of ms, as it will be discussed in Sect. 3.4.2.

Figure 3.16 shows the evolution of NO_x measurements during a cycle with SOI steps, while Fig. 3.17 illustrates that other significant variables (as intake and exhaust pressure, gas temperatures, coolant temperature and air mass flow) do not vary in an important way during the tests. The maximum variation in these properties is 2.8 % for exhaust pressure (considering a moving average filtered value), while the rest of them varies less than 0.5 %.

Sensor Identification

For illustrating the applicability of using data derived from the SOI steps, a dynamic identification is done assuming a linear first order model behaviour (other possible model structures with physical insight could be found in Zhuiykov (2010)) with the transfer equation in the s-domain

$$G(s) = \frac{e^{-s\tau_{NO_x}}}{1 + T_{NO_x} s} \tag{3.1}$$

[3] This delay is in the order of 20 ms maximum for an engine spinning at 3,000 rpm, which can be neglected when comparing with typical delays of the sensor, which are above 500 ms. On the other hand, maximum software delay is below 40 ms, which is not additive to the previous one.

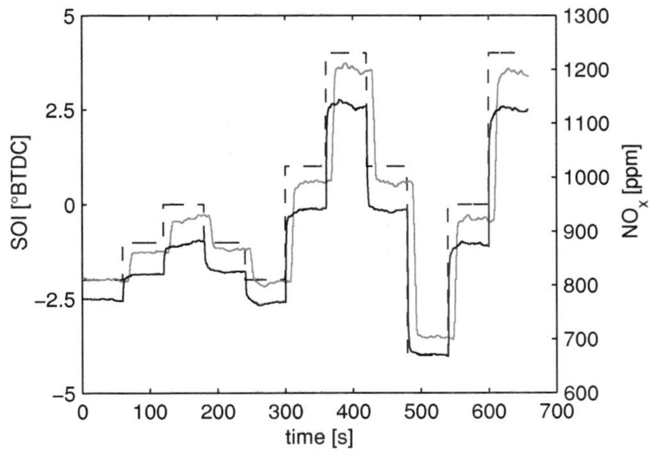

Fig. 3.16 NO_x concentration during a SOI step-like test at 2,000 rpm and 50 mg/stroke. Legend: *dashed line* is SOI, *black line* is the NO_x sensor (z_{NO_x}) before the steady-state calibration and *gray line* is the gas analyser. The difference in terms of delay, dynamics and steady-state values between both sensors is clearly noticed in the Figure

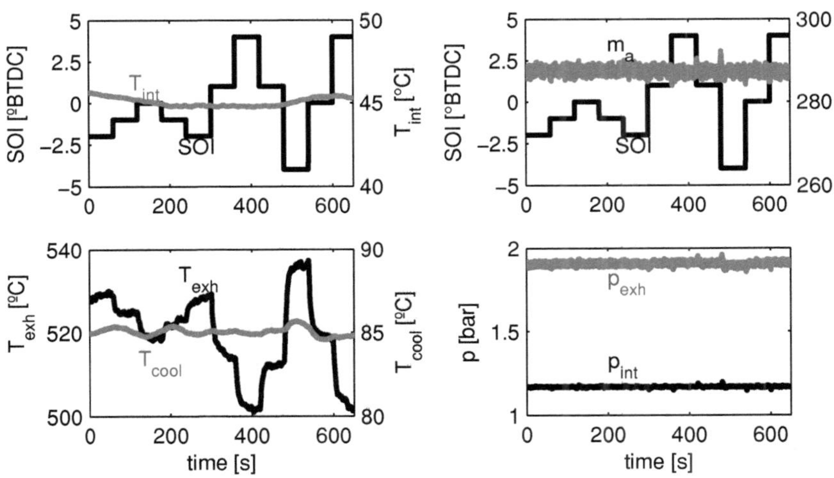

Fig. 3.17 Evolution of different engine variables during a SOI step-like test at 2,000 rpm and 50 mg/str

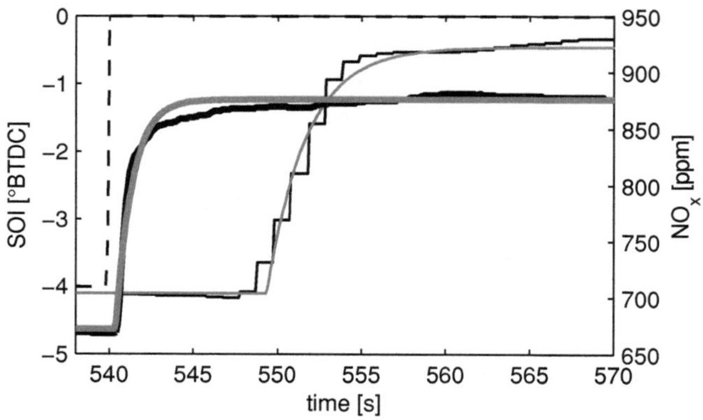

Fig. 3.18 Identified models (*solid gray*) and measured evolution (*solid black*) for both sensors: thin lines correspond to the gas analyser while thick ones to the NO_x sensor. SOI step is reproduced with a black dashed line. The delay of the gas analyser is clear with respect to NO_x sensor

Table 3.8 Mean and standard deviations for the identified time constant T_{NO_x}

	Gas analyser	NO_x sensor
T_{NO_x} (s)	4.25	0.91
σ_k (s)	0.46	0.47

The sensor response to the step is then

$$z_{NO_x}(t) = z_{NO_x}(t_0) + \Delta NO_x \left(1 - e^{-\frac{t-\tau_{NO_x}}{T_{NO_x}}}\right) \qquad t \geq \tau_{NO_x} \qquad (3.2a)$$

$$z_{NO_x}(t) = z_{NO_x}(t_0) \qquad t < \tau_{NO_x} \qquad (3.2b)$$

where $z_{NO_x}(t_0)$ is the z_{NO_x} initial value before the step, ΔNO_x is the step-amplitude and t the time. The sensor delay (τ_{NO_x}) represents the delay from the application of SOI (judged to be instantaneous) to the start of the sensor output variation, and T_{NO_x} stands for the sensor time constant. For each individual step and sensor, T_{NO_x} and τ_{NO_x} are obtained. For the coefficient identification, a least-square fit of (3.2a), (3.2b) is performed [although more robust recursive identification techniques could be used for the online case (Ljung 1999)]. Even though not shown in this work, other different models have been proved, as for instance second order filters and AR or ARMAX models.

Figure 3.18 illustrates an example of the fit obtained with the first order response model. Averaged values of the identified time constant T_{NO_x} and their deviations are shown in Table 3.8. On the other hand, the standard deviations obtained for T_{NO_x} estimates merit for additional analysis, as they suggest that the sensor behaviour could be affected by the gas concentration conditions during the test.

No direct influence is found on the sensors response from the gas speed, temperature or pressure, since the identified parameters do not vary significantly

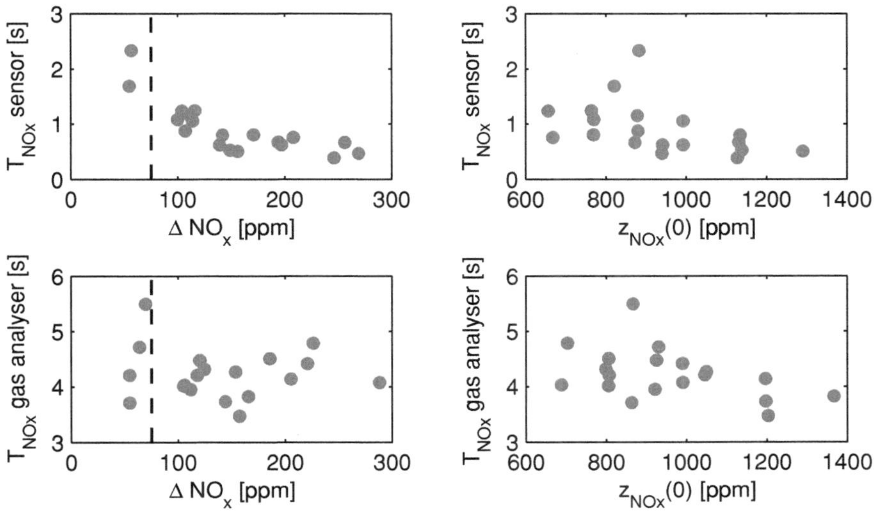

Fig. 3.19 Response time T_{NO_x} for the NO_x sensor as function of the concentration step ΔNO_x and the initial concentration $z_{NO_x}(0)$

with these conditions. In order to analyse the effect of the initial z_{NO_x} concentration ($z_{NO_x}(0)$) and of the step size (ΔNO_x), scatter plots of T_{NO_x} are presented in Fig. 3.19.

In the case of the ZrO_2 sensor, there exists an important variability of T_{NO_x}, especially at low ΔNO_x, with an exponential-like shape. One reason is that, in the case of small steps, the concentration profile is not step-like, but affected by other slower variations (temperature, pressure, etc.) that here are not negligible; another possible reason could be related to the diffusion speed of the gas inside the sensor chambers, which depends on the concentration itself and its evolution along time. Note also that the sensor presents a low sensitivity to low NO_x values Moos et al. (2008). This explains the larger deviations obtained in T_{NO_x} as presented in Table 3.8. However, the response time of the gas analyser does not present these problems and the trend is more constant (and response quite slower) for all values (see bottom plots in Fig. 3.19). For larger step sizes, the response time of the NO_x sensor is more stable, as shown in Table 3.9 where tests with $\Delta NO_x < 75$ ppm have been removed. Physical modelling approaches can be useful for identifying these behaviours by including dynamic models for diffusion chambers Wang (2007).

Furthermore, only low ΔNO_x could be applied in tests with low $z_{NO_x}(0)$, because SOI actuation does not permit significant variations in the gas concentration. Hence $z_{NO_x}(0)$ and ΔNO_x are correlated. This partly explains the effect of high deviation presented for low $z_{NO_x}(0)$ values shown in right-hand plots of Fig. 3.19. In addition, the low sensitivity of ZrO_2 sensors to low NO_x values can be addressed Groß et al. (2012).

Table 3.9 Identified dynamics for NO_x sensors considering tests with $\Delta NO_x > 75$ ppm

	Gas analyser	NO_x sensor
τ_{NO_x} (s)	4.14	0.79
σ_k (s)	0.33	0.27

The discussion about the sensor delay is included in a different subsection. Finally, one of the main problems of automotive sensors is their degradation during their lifespan. For ensuring the correct performance of the control structure, the system has to be capable of checking the sensor periodically. This methodology could be useful for these purposes.

Correspondence with the Discrete Filter

The results above have been presented in the s-domain and the continuous time domain, due to the easy interpretation and for supporting the discussion. However, electronic systems work on the discrete domain. An alternative for converting the obtained results to the z-domain is using the first order discrete filter to model the sensor response

$$G(z) = \frac{1 - a_{NO_x}}{1 - a_{NO_x} z^{-1}} z^{-\tau_{NO_x}/T_s} \tag{3.3}$$

instead of the continuous filter where z is the z-transform variable, a_{NO_x} represents the response time and τ_{NO_x} is the delay. Depending on the sample time (T_s), a_{NO_x} can be deduced Ogata (2001).

Prospective for the On-board Application

Presented method based on shifting SOI could be easily applied during operation phase of automotive diesel engines. That means that the ECU could use this test for identifying the sensor behaviour, and diagnosing the sensor and system performance. This could be periodically done during specific conditions of the engine run. The knowledge of the dynamic sensor characteristics is needed for data fusion and signal reconstruction techniques Manchur and Checkel (2005); Tschanz et al. (2012). For this, the dispersion of the sensor model might be corrected online using the combination of the mentioned techniques and the sensor measurements by using the proposed method. In this way, the robustness of the control law can be checked regarding this dispersion.

Note that the presented results correspond to a NO_x sensor located upstream of the AT devices. Hence the presented methodology will be of direct application for sensors in this position, which can be used for feed-forward control of the SCR system Hofmann et al. (2004). For the case of the sensor located downstream of a SCR system, the following two aspects must be considered:

- The method can be applied but the model should take into account the effect of the SCR. On one hand, the important removal of the raw NO_x concentration (about 80 %) compromises the NO_x output due to the insufficient accuracy of the sensor for low NO_x values. On the other hand, the cross sensitivity to ammonia affect the sensor measurement. Because of these, the dynamic characteristics of the signal

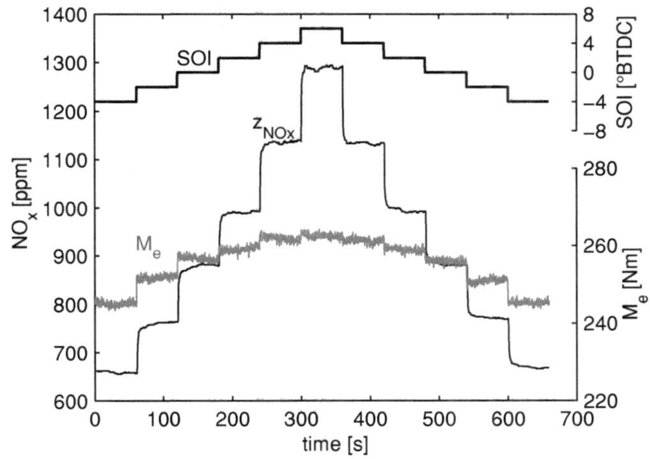

Fig. 3.20 NO_x concentration measured by the NO_x sensor (z_{NO_x}) and torque evolution (M_e) at 2,000 rpm and 50 mg/stroke when varying SOI

can be importantly distorted (also filtering the signal and thus losing the step-like behaviour).

- Even in the case the SCR is not operating, the possibility of exciting the NO_x sensor needs to be checked: depending on the signal dynamics and the volume of AT devices, the gas concentration can sufficiently keep its step-like profile. This issue must be verified beforehand by installing two sensors, one up and one downstream of the AT line. Anyway, once established the feasibility of the method, one of them may be removed. Finally, the NH_3 storage mechanism could also contribute to unexpected variations in the NO_x concentration at the SCR outlet, and hence a sufficient delay is needed between the SCR cut-off and the dynamic characterisation checking.

However, the method could still be used for the SCR system diagnosing, or a proper dynamic model as e.g. the one presented in Hsieh and Wang (2011) might be used for covering the storage dynamics and NH_3 effects.

Excellent operating condition for performing the method is highway operation (where torque and engine speed are reasonably constant during long periods and SOI variations can produce appreciate variations). Figure 3.20 shows the variation in NO_x concentration and torque when SOI is varied in a 2-degree step profile while keeping constant m_f and n. As shown in the central part of the plot, the raw NO_x production can clearly be increased with a low impact on torque; that means that the driveability can be ensured while the test is performed (corrections on m_f can be performed for maintaining the engine torque to its set-point ensuring the driver to be blind to the test). The increase in NO_x emission during the sensor testing should not

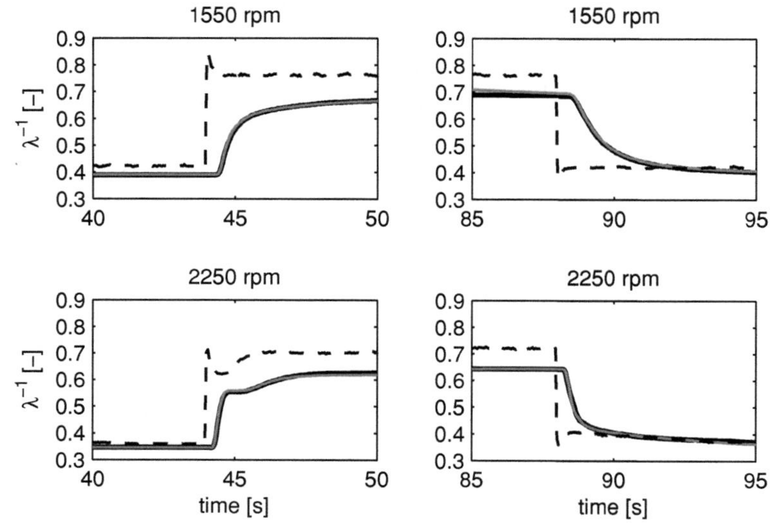

Fig. 3.21 Two sensor responses ($z_{\lambda^{-1}}$ in *solid lines*) for injection steps from 15 to 30 mg/str (*left plots*) and the opposite (*right plots*) at 1,550 rpm (*top plots*) and 2,250 rpm (*bottom plots*). The model x_{λ}^{-1} is provided for comparison (*dashed black*)

importantly impact the overall emissions, since a few seconds are enough for the test (and also a descendent step could be used). On the other hand, in engines with SCR technology and a NO_X sensor located upstream of the SCR, the SCR could cope with the excess of NO_X.

3.4.2.2 Dynamic Calibration of the λ^{-1} Output

The slow dynamic response of λ^{-1} from the NO_X sensor is explained by several reasons: the distance and the system volume between engine exhaust and sensor location affect transport delay and distort the signal; the sensor measurement principle and sensor hardware itself; and the acquisition chain. Although slow, the sensor is quite precise, since several repetitions of the same test provide similar results. This is illustrated in Fig. 3.21, where 2 repetitions of the same fuel step are depicted for 2 different engine speeds. Top plots show a step up (left) and step down (right) in injection for 1,550 rpm and bottom plots the same for 2,250 rpm. Despite the sensor precision, it must be considered that the dynamic response is significantly affected by the operative conditions (e.g. the sensor response in bottom plots in Fig. 3.21 is always faster than in top plots). A model based on the fuel-to-air ratio calculation ($x_{\lambda^{-1}}$) is also shown in the plots (detailed explanation is found in the Chap. 4).

A first order filter similar to that of (3.3) is used and parameters identified on the basis of the data obtained by performing steps on m_f, which provoke step-like

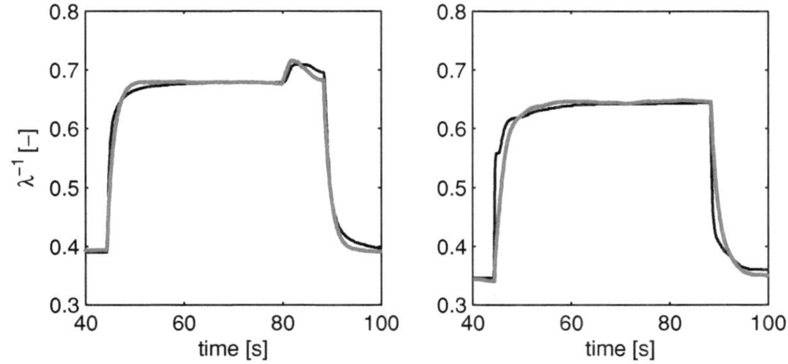

Fig. 3.22 Example of identification of the sensor behaviour for two engine speeds, where $z_{\lambda^{-1}}$ and y_λ^{-1} are depicted. *Top plot* training data at 1,550 rpm where parameters $a_{\lambda^{-1}}$ and $\tau_{\lambda^{-1}}$ are identified. *Bottom plot* validation data at 2,250 rpm, where parameters $a_{\lambda^{-1}}$ and $\tau_{\lambda^{-1}}$ obtained for the top plot situation are used now for obtaining y_λ^{-1}

transitions on λ^{-1} (Guardiola et al. 2013). Figure 3.22 shows the evolution of the measured $z_{\lambda^{-1}}$ and the identified sensor model y_λ^{-1} for two different operating conditions. Left plot is used for identification while validation is provided by the right plot. The fitting is good (even in the right plot) although a slight error can be appreciated, which suggests that the dependence of the sensor dynamics with the engine operating point may be neglected in a first approach, i.e. the response time may be taken as constant. A similar study to that of the NO_x output may be performed in order to obtain characteristic values of the response time. Anyway, the repetitiveness of the parameters is higher than for the NO_x case, and in this work the NO_x sensor response time for λ^{-1} output is considered constant.

This method could be applied when driver performs a sharp tip-on or can profit DPF regeneration modes, even though the effect of post-injections should be studied more deeply.

3.4.2.3 Delay in the NO_x Sensor

As expected, the NO_x sensor delay is much lower than chemiluminescence system delay. This occurs because the gas analyser delay is mainly caused by the treatment operations and by the transport in the long gas line feeding the measurement system (12 m), which adds a non-negligible delay due to the low flow allowed. The outputs have been plotted in Fig. 3.16, and the significant lower delay for the ZrO_2 sensors with respect to the gas analyser can be clearly noticed.

In principle, the ZrO_2 sensor delay (for both λ^{-1} and NO_x outputs) varies with the sensor actuation possibly by gas concentrations (Wang 2007) and in a minor way by transport lag. The effect of the gas concentrations and mass flows themselves

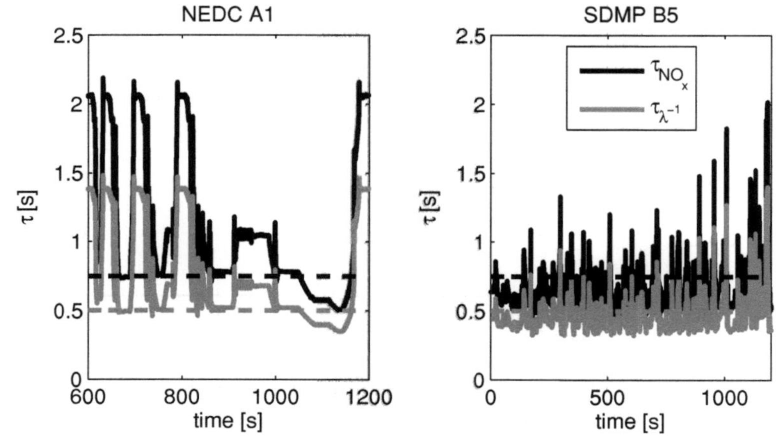

Fig. 3.23 Variable sensor delay modelled for z_{NO_x} and $z_{\lambda^{-1}}$ outputs from the NO_x sensor during the second half of the NEDC A01 (*left*) and SDMP B5 (*right*). Higher peaks correspond to the idling operating conditions, which are overestimated. In general, the bigger m_f and n, the lower the delay

over the final delay is appreciable and should be analysed but it escapes from the scope of this work. With respect to the transport delay, it could be calculated taken into account the gas flow rate and the exhaust pipe cross section. The transport delay can take a wide range of values depending on the operating conditions. Simulations with a wave action model (Galindo et al. 2009) and for the considered diesel engine after varying injected fuel mass and engine speed show velocities of the exhaust flow between 10 m/s and 140 m/s. The transport delay could be then calculated as the time that the gas takes to reach the sensor from the combustion chamber, and goes with the inverse of the gas velocity. From the simulations, transport delays rarely present values above 100 ms, and according to simulations by wave action models, are generally lower than 25 ms. The influence on the total delay is supposed to be minimum with respect to the hardware delay.

Some authors have proposed functions for modelling the delay. For instance, (Trimboli et al. 2012) design the following function

$$\tau(n, m_a) = \alpha_0 + \frac{\alpha_1}{n} + \frac{\alpha_2}{m_a} \tag{3.4}$$

Figure 3.23 shows the delay modelled by (3.4) in the cycles A1 and B5 for both NO_x and λ^{-1} outputs. The results show that the delay is quite constant during part and full load operation, while it is slightly bigger when approaching to idling or low loads. The variation can be explained not really by the (minimum) transport lag but mainly by the sensor diffusion processes between the sensor chambers, which are affected by the total oxygen and NO_x concentrations. In the case of NO_x, the delay is slightly higher than for λ^{-1} due to the sensing element location is in the inner cavity. It should also be noted that the delay tends to be overestimated during idling (about 2 s in NO_x against 1.4 in the reality) due to the simple model structure (3.4). In this work, a constant delay is used for the implementation of models and fusion strategies, and values are shown in dashed lines in Fig. 3.23 (about 0.75 s in NO_x and 0.5 s in λ^{-1}, in line with the manufacturer technical data). The results obtained in subsequent chapters justify this decision, in addition to the simple programming of a constant delay against a variable one. Furthermore, the robustness, dispersion and ageing of a delay model should be assessed more in detail, where scheduling strategies in line with the thesis algorithms might be explored.

3.5 Conclusions

This Chapter has described the experimental setup and the test campaign performed in order to design and validate the models and algorithms utilised in the current dissertation. Furthermore and considering the gas concentration sensors used for this work, the NO_x and λ^{-1} outputs from a ZrO_2 sensor are characterised by means of steady-state tests for static calibration and transient steps for dynamic characterisation. More precisely:

- A simple on-engine experiment for characterising NO_x output is presented. The method uses the engine as a gas generator and the procedure consists of applying steps in SOI producing sharp variations in NO_x concentration. With a simple variation of 2–3° in SOI, NO_x concentration is varied whilst torque values and fluid conditions remain approximately constants.
- A similar procedure based on performing injection fuel rate steps in order to produce sharp variations on λ^{-1} is used for the dynamic characterisation of this signal.

These methods allow the determination of the dynamic response of the sensor (for both NO_x and λ^{-1}) over its entire useful life and the detection of faults in the sensor. A simple mathematical modelling, taking as a reference a first order model response, is proposed, although more complex models might be utilised. According to this, the delay and the response time is modelled for both NO_x and λ^{-1}. In both cases, and even though there exists variability in the parameters, the delay and the response time are considered constants for the applications made in subsequent chapters. The good results of the models and fusion strategies for the proposed applications give validity to this decision, provided that a deeper analysis is advised for future robust implementations. Due to its nature, the methods could be implemented on-board for the online dynamic calibration and development of adaptive strategies.

References

Arrègle J, Bermúdez V, Serrano JR, Fuentes E (2006) Procedure for engine transient cycle emissions testing in real time. Exp Therm Fluid Sci 30(5):485–96

Bermúdez V, Luján JM, Serrano JR, Pla B (2008) Transient particle emission measurement with optical techniques. Measur Sci Technol 19(6):065404

Broatch A, Luján JM, Serrano JR, Pla B (2008) A procedure to reduce pollutant gases from diesel combustion during European MVEG-A cycle by using electrical intake air-heaters. Fuel 87(12):2760–78

Chan SH, Chen XS, Arcoumanis C (1997) Measurement and signal reconstruction of transient nitric oxide emissions in the exhaust of a turbocharged diesel engine. J Dyn Syst Measur Control Trans ASME 119(4):620–0

Galindo J, JM Luján H Climent, Guardiola C (2007) Turbocharging system design of a sequentially turbocharged diesel engine by means of a wave action model. SAE Technical Paper 2007–01-1564

Galindo J, Serrano JR, Arnau FJ, Piqueras P (2009) Description of a semi-independent time discretization methodology for a one-dimensional gas dynamics model. J Eng Gas Turbines Power 131(3):034504

Galindo J, Serrano JR, Guardiola C, Blanco-Rodriguez D, Cuadrado IG (2011) An on-engine method for dynamic characterisation of NO_x concentration sensors. Exp Therm Fluid Sci 35(3):470–6

Geivanidis S, Samaras Z (2008) Development of a dynamic model for the reconstruction of tailpipe emissions from measurements on a constant volume sampling dilution system. Measur Sci Technol 19(1):015404

Groß A, Beulertz G, Marr I, Kubinski DJ, Visser JH, Moos R (2012) Dual mode NO_x sensor: measuring both the accumulated amount and instantaneous level at low concentrations. Sensors (Basel) 12(3):2831–50

Guardiola C, Pla B, Blanco-Rodriguez D, Mazer A, Hayat O (2013) A bias correction method for fast fuel-to-air ratio estimation in diesel engines. Proc Inst Mech Eng Part D J Automob Eng 227(8):1099–111

Hofmann LK, Rusch S, Fischer, Lemire B (2004) Onboard emissions monitoring on a HD truck with an SCR system using NO_x sensors. SAE Technical Paper 2004–01-1290

HORIBA (2001) Horiba MEXA-7000DEGR instruction manual

Hsieh M-F, Wang J (2011) Design and experimental validation of an extended Kalman Filter-based NO_x concentration estimator in selective catalytic reduction system applications. Control Eng Pract 19(4):346–53

Ljung L (1999) System identification: theory for the user. Prentice Hall PTR, Upper Saddle River

Luján JM, Galindo J, Serrano JR, Pla B (2008) A methodology to identify the intake charge cylinder-to-cylinder distribution in turbocharged direct injection diesel engines. Measur Sci Technol 19(6):065401

Manchur TB, Checkel MD (2005) Time resolution effects on accuracy of real-time NO_x emissions measurements. SAE Technical Paper 2005–01-0674

Moos R, Spörl M, Hagen G, Gollwitzer A, Wedemann M, Fischerauer G (2008) TWC: lambda control and OBD without lambda probe—an initial approach. SAE Technical Paper 2008–01-0916

Mrosek M, Sequenz H, Isermann R (2011) Identification of emission measurement dynamics for diesel engines. In: IFAC proceedings volumes (IFAC-PapersOnline), vol 18, pp 11839–11844

Ogata K (2001) Modern control engineering, 4th edn. Prentice Hall, Upper Saddle River

Payri F, Arrègle J, López JJ, Mocholí E (2008) Diesel NO_x modeling with a reduction mechanism for the initial NO_x coming from EGR or re-entrained burned gases. SAE Technical Paper 2008–01-1188

Regitz S, Collings N (2008) Fast response air-to-fuel ratio measurements using a novel device based on a wide band lambda sensor. Measur Sci Technol 19(7):075201

Trimboli S, Di Cairano S, Bemporad A, Kolmanovsky IV (2012) Model predictive control with delay compensation for air-to-fuel ratio control, vol 423. Lecture notes in control and information sciences, Springer, Berlin

Tschanz F, Amstutz A, Onder CH, Guzzella L (2012) Feedback control of particulate matter and nitrogen oxide emissions in diesel engines. Control Engineering Practice (in Press)

Wang DY (2007) Real-time dynamics of amperometric exhaust oxygen sensors. Sens Actuators B: Chem 126(2):551–556

Zhuiykov S (2010) Electrochemistry of zirconia cas sensors. Taylor & Francis, London

Chapter 4
Control Models for Engine-Out NO_x and λ^{-1}

4.1 Introduction

Control oriented modelling is aimed to predict engine variable with a low computational effort and is the basis of model based strategies, such as diagnosis or MPC (Deng et al. 2012). Models usually have the possibility of predicting with acceptable accuracy the transient performance of the engine. Once that the Chap. 2 have already reviewed models for NO_x and λ^{-1}, this chapter is devoted to the design of an ECU signals based virtual sensor for the estimation of engine-out λ^{-1} and NO_x (Payri et al. 2012a, b; Guardiola et al. 2013). The reported results correspond to the DW12B DICI engine presented in the Chap. 3 with all AT devices removed.

4.2 λ^{-1} Model

Instead to refer to λ, results and discussion are centred in its inverse λ^{-1}, namely relative fuel-to-air ratio or simply richness, as it presents a bounded value (ranging from 0 to 1 in diesel engines, although slightly higher values might be reached during engine load transients).

Depending on the engine configuration, an air mass flow sensor ($\dot{m}_a$) may be present (or mass flow can be estimated through the boost pressure), and the injected fuel mass flow ($\dot{m}_f$) is usually estimated by the ECU. A simple model may be then proposed

$$x_{\lambda}^{-1} = 14.5\frac{\dot{m}_f}{\dot{m}_a} \tag{4.1}$$

where $\dot{m}_f$ and $\dot{m}_a$ quantities are expressed in the same units, i.e. mg/str or kg/s, and 14.5 stands for the stoichiometric air-to-fuel ratio. Note that the model neglects the mass accumulation in the intake and exhaust manifolds [the reader is referred to the Chap. 2 for finding more elaborated models]. Despite such effects should be

D. Blanco-Rodriguez, *Modelling and Observation of Exhaust Gas Concentrations for Diesel Engine Control*, Springer Theses, DOI: 10.1007/978-3-319-06737-7_4,

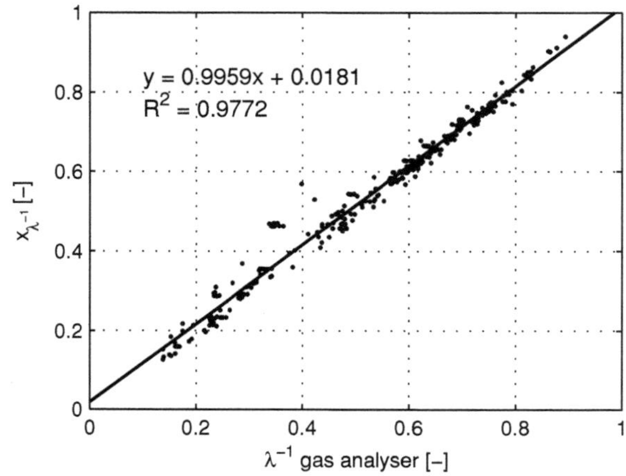

Fig. 4.1 x_λ^{-1} and λ^{-1} from the gas analyser for different steady operating conditions; the linear fit and correlation coefficient are represented

considered for correcting $\dot{m}_a$ during the engine transients (for example, when boost pressure is increased or when shifting between boosting modes Galindo et al. (2009)), they will be neglected hereinafter for simplicity.

Figure 4.1 compares model steady-state results (x_λ^{-1}) with gas analyser measurements. As it can be easily noticed, the model provides a lower accuracy than that of the sensor (shown in Fig. 3.15). Regarding the transient behaviour, the model exhibits an almost instantaneous response; the injected fuel mass estimate, although biased is fast; and the air mass flow measurement characteristic time is in the order of miliseconds, while the exhaust gas sensor response time can be of several hundreds and even seconds (depending on the considered set-up).

4.2.1 λ^{-1} Model Drift

x_λ^{-1} output presents a significant bias that strongly depends on the operating conditions due to several facts:

- m_f is obtained from internal calculations of the ECU, and response is fast and non-delayed, but is based on tabulated values which rely on rail pressure measurement and injection duration. The error of these tables can be significant when short injections or split injection strategies are applied. On the other hand, injector manufacturing discrepancies and ageing can create a significant unit-to-unit [and cylinder-to-cylinder (Macián et al. 2006a, b; Payri et al. 2006)] dispersion in the actual injected quantity.
- While $\dot{m}_a$ sensor response is judged to be fast and non-delayed enough for being directly used as a model input, the precision of the sensor is not very high [for example in (Desantes et al. 2010) the non-systematic error of the air mass flow sensor is quantified with a standard deviation of 3.23 %, in part due to flow pulsations in the intake line associated with the engine speed]. On the other hand,

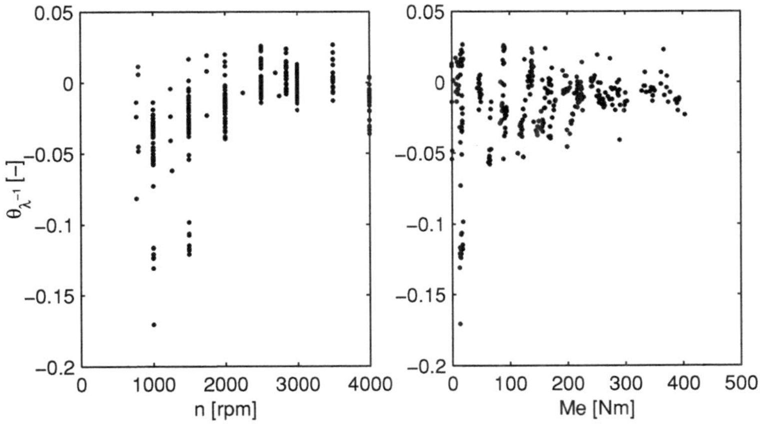

Fig. 4.2 Model bias for the steady-state tests

during transient processes in which a significant variation of the air mass accumulated in the manifolds exists, the air mass flow measured in the intake line is different to that entering the engine (Galindo et al. 2009). This can be corrected accounting pressure variations in the $\dot{m}_a$ estimation, although this is not considered in the current work.

With this scenery, $x_{\lambda^{-1}}$ can be related to the actual value of λ^{-1} considering a bias $\theta_{\lambda^{-1}}$

$$x_{\lambda^{-1}} = \lambda^{-1} - \theta_{\lambda^{-1}} \tag{4.2}$$

The model bias is not constant and varies with the operating condition, since $\dot{m}_a$ error significantly depends on the engine speed, and $\dot{m}_f$ error is affected by the injection profile, which is usually scheduled as a function of n and M_e. Figure 4.2 shows bias estimation for the measured steady points and its variation with engine speed and torque.

Additionally, the model bias varies with time due to the system drift, and a slow variation could be expected. At the same time, the model drift may be associated with injection soiling, air mass flow sensor drift, or the variations of leakage in the intake manifold and blow-by. In order to model that, both effects, the dependency with the operative conditions, and the drift, must be considered

$$\frac{\mathrm{d}\theta_{\lambda^{-1}}}{\mathrm{d}t} = \frac{\partial\theta_{\lambda^{-1}}}{\partial t} + \frac{\partial\theta_{\lambda^{-1}}}{\partial n}\frac{\mathrm{d}n}{\mathrm{d}t} + \frac{\partial\theta_{\lambda^{-1}}}{\partial m_f}\frac{\mathrm{d}m_f}{\mathrm{d}t} + \cdots \tag{4.3}$$

Although the bias variation associated with the system drift ($\partial\theta_{\lambda^{-1}}/\partial t$) is expected to be slow, the actual variation of the bias may be very fast, due to the ability of the engine of performing fast transition between operating conditions (defined by n and m_f).

4.3 NO_x Model

A RT gray box NO_x model combining first principle equations with the use of operating-point dependent look-up tables is developed based on simple correlations of engine available signals. This model describes a quasi-static representation of the engine plus some filters for describing actuators and sensor dynamics, where the scheduling points are n and m_f.

4.3.1 *Previous Considerations*

Firstly, previous considerations related with the validity of a quasi-static representation and definition of engine operating points, the $[O_2]_{int}$ estimation and the selection of the appropriate units for modelling NO_x are discussed.

4.3.1.1 Comments on the Engine Dynamics and Definition of the Engine Operating Point

The engine works in closed loop by mainly following set-point references for air mass flow ($\dot{m}_a^*$) and boost pressure (p_{boost}^*), which are controlled by modifying the actuators for EGR (u_{egr}) and waste-gate (u_{wg}). Other variables such as the coolant temperature (T_{cool}) are measured and taken into account in the ECU controller. Speed (n) and injection (m_f) are used as scheduling variables in order to build the references for the controllers by means of look-up tables. Since the fuel path representative times are in the order of μs, this can be considered as instantaneous if compared with the air path dynamics. However, the major influence of the fuel path over the air path is with the smoke limiter, which limits the injected fuel rate until $\dot{m}_a$ or p_{boost} (depending on the manufacturer) reach a certain limit in order to bound λ^{-1}. Therefore, the engine states (X) could be modelled with the dynamic equation

$$\dot{X} = f(X, n, m_f, u_{egr}, u_{wg}, T_{cool}) \tag{4.4}$$

The set-point references for air and boost pressure ($\dot{m}_a^*$ and p_{boost}^*) are modelled by look-up tables scheduled by n and m_f

$$\dot{m}_a^* = f(n, m_f) \tag{4.5a}$$

$$p_{boost}^* = f(n, m_f) \tag{4.5b}$$

The variables $\dot{m}_a$ and p_{boost} are measured by sensors in diesel production engines, and if considering them as fast signals and neglecting mass storage effects (MAF sensor for measuring $\dot{m}_a$ is usually installed upstream of the low pressure compressor), the dependency of the command signals u_{egr} and u_{wg} might be skipped as follows

$$\dot{X} = f(X, n, m_f, \dot{m}_a, p_{boost}, T_{cool}) \tag{4.6}$$

If the engine air path states can be modelled by quasi-static representations (dynamics do not have influence on the output) or by adding filters and delays function only of the scheduling inputs n and m_f, then the dependency with $\dot{m}_a$ and p_{boost} might also be modelled as function of n and m_f. The case of temperatures, associated with slow transients is more difficult and the measured engine variable T_{cool} is not able to exactly represent for instance in-cylinder conditions. The inner wall cylinder temperature influence deserves a specific consideration, and its influence is included in the model with a dynamic factor (see discussion in Sect. 4.3). Anyway, T_{cool} is used as input for switching warm-cold engine strategies. Based upon this reasoning, X could be represented as

$$\dot{X} = f(X, n, m_f, T_{cool}) \tag{4.7}$$

Putting aside the thermal dynamic effects, further simplifications could be made if the engine performance is slow enough that $\dot{m}_a$ and p_{boost} are able to track $\dot{m}_a^*$ and p_{boost}^* fast and then, engine performance is simplified to

$$X = f(n, m_f) \tag{4.8}$$

Figure 4.3 shows examples of the set-point references and responses for the air and fuel path and for three different cycles (NEDC A1, CADC B2 and SDMP B5), including also in the bottom plots the variables u_{egr} and u_{wg} corresponding to the actuation signals for EGR and WG respectively. The cycles correspond in that order from the slowest to the fastest transitions. In the A1 test, the air path controllers track with a minimum error both $\dot{m}_a^*$ and p_{boost}^*, especially for the case of $\dot{m}_a$ which is logical due to the slower dynamics associated to the turbo inertia. The response for the B2 cycle is also good but transients are a little bit more aggressive than for A1, and the errors are slightly higher. For these two cycles, the quasi-static representation defined by (4.8) gives good results if the engine is already warmed up (T_{cool} is stable). However, for the case of B5 cycle, the transients are much more aggressive and errors on the $\dot{m}_a$ and p_{boost} tracking are appreciable as well as the fact that the EGR valve is mostly closed during the cycle and the second TC is indeed working at some points (see Fig. 4.3 when u_{wg} reaches 100 %). For this cycle, the engine is running out of the nominal ($\dot{m}_a^*$ and p_{boost}^* set-points differ from the other cycles) and the effect of the thermal loading due to the fast transients is critical.

Thus, the selection of the model structure for a given precision would depend not only on the engine characteristics but on the use cases. In all these cases and examples, it should be considered not only the thermal transient effects but the sensor dynamics. Therefore, (4.8) should consider the sensor model if the output is compared with a sensor output. The model structure proposed in this chapter, aimed to be representative of the engine performance under different conditions, is built on the basis of (4.6) as a quasi-static model representation and an extra-state for representing sensor dynamics (now included in the equation):

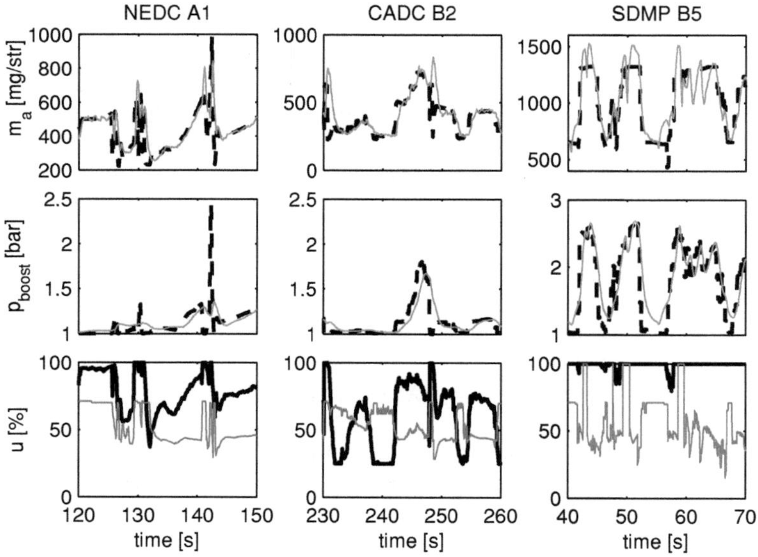

Fig. 4.3 Air and fuel path response to variations in the driver's command (u_α) for different cycles. First and second row plots: *dashed line* is set-point $\dot{m}_a^*$ and p_{boost}^*, and *gray line* is measured value of $\dot{m}_a$ and p_{boost}. Last row plots: *gray line* u_{wg} [mg/str], *black line* u_{egr} [mg/str] is the smoke limit for injection. Note that 100 % in u_{egr} corresponds to the EGR valve fully closed and around 25 % the EGR mass flow is maximum (the relation between u_{egr} command and the effective area is not linear). For the case of u_{wg}, 100 % corresponds to WG fully closed and the two turbo shafts working

$$X = f(n, m_f, \dot{m}_a, p_{boost}, T_{cool}) \tag{4.9a}$$

$$\dot{X} = f(X, t - \tau) \tag{4.9b}$$

where t is the time and τ the sensor delay. This model structure, considering an state for including the thermal transient effect, is used after along the present chapter for developing a NO_x model. Alternatively, in Sect. 6.3.1 a similar model to (4.8) will be considered for modelling NO_x in engine cycles.

4.3.1.2 Intake Oxygen Estimation

In the interest of developing a RT model for NO_x, the Arrhenius equation (2.6) can be arranged with two operating point dependent factors C_1 and C_2

$$\frac{dNO_x}{dt} = C_1 e^{-C_2/T} \tag{4.10}$$

where C_1 includes gas intake concentrations (nitrogen and oxygen) while C_2 is the exponential factor. According to Andersson (2006), oxygen rate in the cylinder and

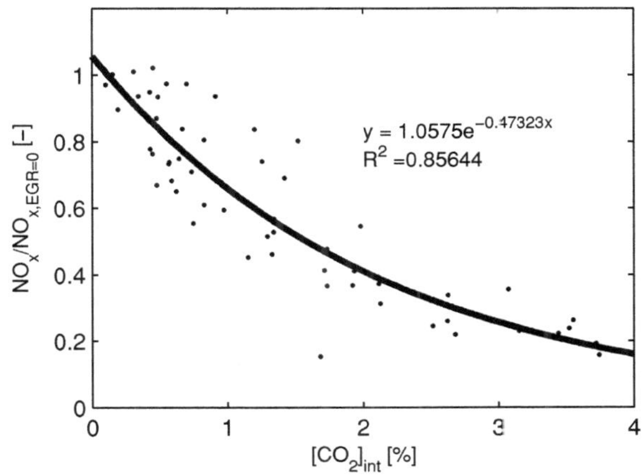

Fig. 4.4 Normalised NO_x emissions as function of $[CO_2]_{int}$ for different nominal operating points in the diesel engine explained in Chap. 3 and coming from the steady-state campaign. Engine calibration is the standard given by the manufacturer

operating conditions (n and m_f as discussed above) seem to be suitable variables for predicting NO_x emissions. Hence it is possible to find a new generic expression whose structure is equivalent to (4.10):

$$x_{NO_x} = f([O_2]_{int}, x_1, ..., x_n) \tag{4.11}$$

where x_{NO_x} is the model output. This expression relates NO_x with $[O_2]_{int}$ and different parameters ($x_1, ..., x_n$) including temperature and time reactions effects. The model, which is based on the thermal NO_x formation mechanism, might be calibrated for accounting for the NO_x formation by other mechanism.

Figure 4.4 shows NO_x versus $[CO_2]_{int}$ in the intake manifold when varying EGR valve position for different speed n and injected fuel rate m_f. NO_x is normalised by $NO_{x,EGR=0}$, which corresponds to the maximum NO_x value at every operating point (equivalent to the EGR valve fully closed). The exponential fit in Fig. 4.4 is clear, in spite of operating point dependency of the exponential factor is not modelled yet.

$[O_2]_{int}$ can be measured, but intake lambda probes are not always available in diesel engine production cars, and measurements errors related with the gas non-homogeneity may appear (Luján et al. 2008; Arrègle et al. 2010). Considering the test engine, Sect. 3.2.1 already compared the intake oxygen estimation by a UEGO sensor with different samples in different intake positions measured by a gas analyser. Although the steady-state accuracy is acceptable, the UEGO sensor presents important problems linked with the pressure effects over the output signal, which drive to saturation and the need of compensation.

Alternatively, $[O_2]_{int}$ can be modelled by relating $[O_2]_{int}$, EGR and λ^{-1}

$$[O_2]_{air}(1 - \lambda^{-1}) = [O_2]_{exh}. \tag{4.12}$$

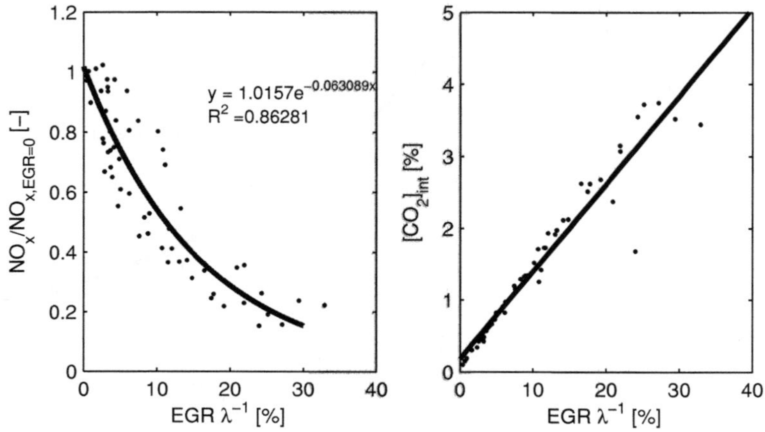

Fig. 4.5 *Left plot* shows normalised NO_x emissions as function of $EGR\lambda^{-1}$ for different operating points in the diesel engine explained in Chap. 3. *Right plot* shows the fitting of $[CO_2]_{int}$ and $EGR\lambda^{-1}$

By the oxygen balance at the intake junction

$$[O_2]_{int} = [O_2]_{air}(1 - EGR) + [O_2]_{exh}EGR \tag{4.13}$$

and substituting the former in the latter

$$[O_2]_{int} = [O_2]_{air}(1 - EGR\lambda^{-1}) \tag{4.14}$$

that shows a direct relationship between $[O_2]_{int}$ and $EGR\lambda^{-1}$. The product $EGR\lambda^{-1}$ is often called the inert gas rate and represents the actual portion of exhaust gas that has reacted with the injected fuel and then contains no available oxygen; e.g Andersson et al. (2006) use $EGR\lambda^{-1}$ for designing a fast NO_x model. Figure 4.5 shows the NO_x fitting with $EGR\lambda^{-1}$ and the fitting between $EGR\lambda^{-1}$ and $[CO_2]_{int}$, where the expression (4.14) is proved.

On the other hand, Fig. 4.6 shows the fitting of NO_x by using *EGR*. Results with *EGR* are much worse than with $EGR\lambda^{-1}$ and it is difficult finding a tendency, which proves that $EGR\lambda^{-1}$ seems a good alternative for estimating $[O_2]_{int}$ (or $[CO_2]_{int}$) when an intake lambda probe is not installed, although the uncertainties propagated from the different involved measurements are to be checked (Arrègle et al. 2010).

4.3.1.3 About the NO_x Model Output Units

Sensor raw signal is measured in ppm, which indicates the relative molar concentration of NO_x, but the final NO_x mass depends on the exhaust mass flow and time. The conversion from ppm to g/h used in this work is

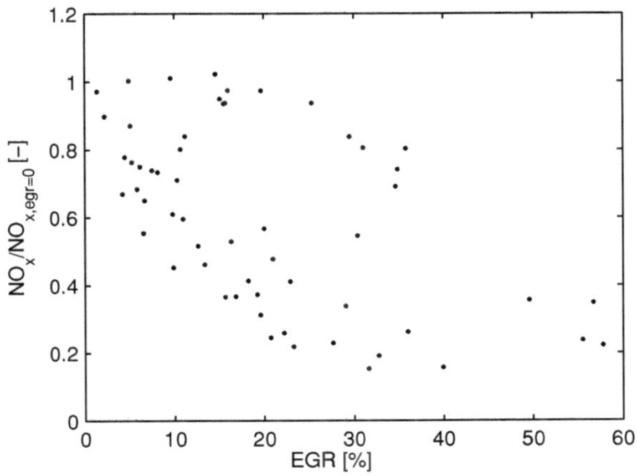

Fig. 4.6 Normalised NO_x emissions as function of *EGR* for different nominal operating points of the DW12B engine

$$\dot{x}^{g/h}_{NO_x} = 0.001587 x^{ppm}_{NO_x} \cdot (\dot{m}^{kg/h}_a + \dot{m}^{kg/h}_f) \tag{4.15}$$

considering that molecular mass of NO_x is 46.1 g/mol and exhaust mass flow density is 1.293 kg/m^3, measured at 273 K and 101.3 KPa. The conversion of mg/str to kg/h is defined as

$$\dot{m}^{kg/h} = 1.2\,10^{-4}\, n \cdot m^{mg/str} \tag{4.16}$$

in a 4 cylinder basis. The use of mass flow variables is appropriate for taking into account total emissions and calculating the real impact on the environment. In addition, AT devices such as the LNT require a count of the accumulated NO_x mass for regenerating the trap. Then, a mass flow seems also suitable for OBD purposes.

For computing total NO_x emissions, $\dot{x}^{g/h}_{NO_x}$ can be integrated in the considered cycle

$$M^{g}_{NO_x} = \int_{t_1}^{t_2} \dot{x}^{g/h}_{NO_x}\, dt \tag{4.17}$$

where *M* stands for total mass.

Commercial NO_x sensors measure the relative volume flow of NO_x in parts per million (ppm) and the conversion to g/h can induce some error, especially if $\dot{m}_a$ presents some drift or filtering effect. Anyway, the use of g/h for online NO_x tracking is justified for two reasons: (1) dilution effects are avoided as ppm is a relative measurement and (2) the use of g/h is coincident with the actual emissions. Dilution effects can distort the ppm output and can lead to sign shifts in the gradient when comparing ppm output and g/h at high load conditions (no EGR), i.e. an increase of g/h could result in a decrease of ppm if $\dot{m}_a$ increases. On the other hand, minimum variations of g/h can lead to huge variations in ppm even though the relative gradient could be high in both two, especially at low load conditions (related with low $\dot{m}_a$), leading to overestimate NO_x emissions when tracking ppm. In this work, the NO_x

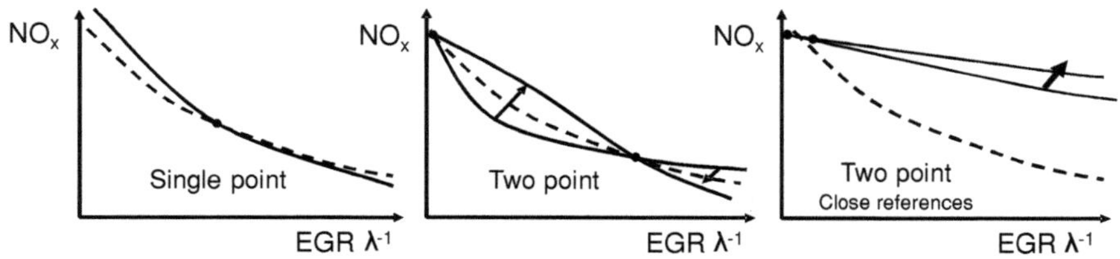

Fig. 4.7 Single point reference model (*left plot*) versus two-point reference models (*center* and *right plots*)

model output is calibrated for minimising the error on g/h, even though ppm output is also considered.

4.3.1.4 Selection of the Model Structure

In order to find a suitable model equation for a RT NO_x model, different structures may be used (Finesso and Spessa 2012; Lee et al. 2013). Figure 4.7 shows a visual example of point dependent fittings: single (left plot) and two-point functions (centre and right plot). On one hand, if using a single point function, model variations are centred around one reference. The point should be chosen as representative of the engine operation: NO_x emissions under nominal conditions (defined with the standard calibration of the OEM) or maximum emissions ($EGR = 0$), which is used in previous figures. On the other hand, the two point function interpolates in a two reference basis, which in diesel engines could be around nominal EGR actuation and closed EGR (at least for the EGR area of the engine).

Both correlations have advantages and drawbacks: single point fitting is more flexible than two point for mathematical adjustment but with a smaller accuracy in points far from the nominal, however the single point one is easier to calibrate as only one map or reference must be tuned. Admittedly, extrapolation capability for the two-point function output when input values are out of the two-point area is compromised. See for instance right plot of Fig. 4.7, the two references are quite close to each other and extrapolation is not fair. The same occurs in diesel engines in areas where nominal $EGR\lambda^{-1}$ is low. Here, the single reference point model is used due to simplicity and robustness.

4.3.2 A Real Time NO_x Model

A RT NO_x model is proposed using $EGR\lambda^{-1}$ for representing the intake oxygen jointly with maps and corrections fitting the operating point dependency (Finesso and Spessa 2012). The model scheme is outlined in Fig. 4.8. In the top right and for clarifying the reader, engine positions subscripts are reminded: *boost* is downstream the intercooler, *int* the intake manifold, *egr* the EGR manifold at the

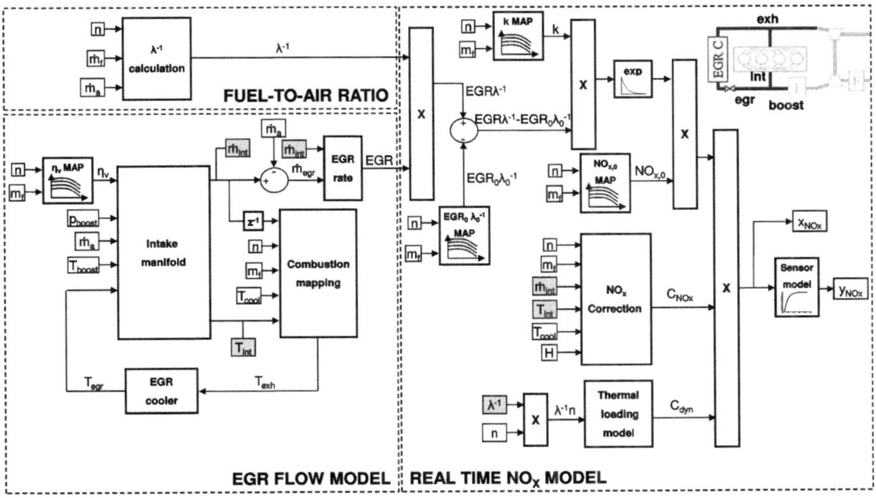

Fig. 4.8 Structure of the RT NO_X model, where the main blocks are depicted: the EGR flow model (or air path model), the fuel-to-air ratio calculation and the proper NO_X model calculation. The model inputs (coming from ECU signals) are marked with *squares with white front*, while the *squares with gray front* are used as short-cuts for linking intermediate variables in the model. At the *top right*, the legend for the engine positions is shown

intake junction and *exh* the exhaust. The main RT NO_X model requires two additional blocks: a mean value engine calculation of the EGR flow for obtaining *EGR* and the calculation of the fuel-to-air ratio λ^{-1}. The model design is flexible and other structures for the EGR flow and λ^{-1} model or other measurements might be utilised just replacing the corresponding blocks.

Following the discussion presented in the previous section about representation of engine states, an engine-out NO_X model is designed considering the state vector X

$$X = \begin{bmatrix} x_{NO_X} & y_{NO_X} & C_{dyn} \end{bmatrix}^T \tag{4.18}$$

where x_{NO_X} is the NO_X model output (actual NO_X), y_{NO_X} is the filtered and delayed NO_X output (for comparing with NO_X sensor) and C_{dyn} is a dynamic factor for coping with in-cylinder temperature and explained after. The input signals contained in the vector U are available in the ECU and are shown in squares in the Fig. 4.8

$$U = \begin{bmatrix} n & m_f & \dot{m}_a & p_{boost} & T_{boost} & T_{cool} & H \end{bmatrix}^T \tag{4.19}$$

where H is the humidity and the rest of variables have already been defined. Model output vector Y is the modelled sensor response

$$Y = y_{NO_X} \tag{4.20}$$

in order to have a comparable signal with NO_x sensor output during dynamic tests. x_{NO_x} indicates the raw or actual NO_x, which must be filtered and delayed. The delayed first order discrete model identified in Chap. 3 is used

$$y_{NO_x} = z^{-\tau_{NO_x}/T_s} \frac{1 - a_{NO_x}}{1 - a_{NO_x} z^{-1}} x_{NO_x} \tag{4.21}$$

where a_{NO_x} is the sensor response time and τ_{NO_x} is the total sensor delay (see Chap. 3 for more information).

Finally, constants are also marked with squares and gray squared variables are used as shortcut for avoiding crossed lines in the Fig. 4.8.

4.3.2.1 Model Hypotheses

The following hypotheses are assumed:

- The model is programmed and calibrated in discrete form and frequency used for simulation is 50 Hz.
- The model is based on a set of static look-up tables, curves and parameters with the addition of discrete filters and lag blocks in order to consider system dynamics.
- Intake conditions ($\dot{m}_{int}$ and T_{int}) are calculated on the basis of the volumetric efficiency and an EGR flow model by assuming constant pressures upstream and downstream of the intake junction and an adiabatic mixing model for the temperature.
- The engine combustion is tabulated by using look-up tables related to nominal conditions and different variations of the intake conditions.
- The EGR path is modelled with a cooler model and a first order filter for accounting with the manifold dynamics.
- Turbocharger effects are bypassed by directly using sensor signals $\dot{m}_a$ and p_{boost}, the former located upstream of the compressor and the latter downstream of the intercooler.

In the next, the model and sub-model equations are described.

4.3.2.2 RT NO_x Model Equation

The NO_x model is presented in (4.22) and consists in a single function referenced to conditions with nominal EGR for calculating x_{NO_x}, which is affected by an exponential $EGR\lambda^{-1}$ correction and additional factors C_{NO_x} and C_{dyn}. This structure guarantees that nominal tests corrections are eliminated minimising the nominal error and subscript 0 goes with nominal conditions. Then, the model x_{NO_x} output is calculated as follows

$$x_{NO_x} = NO_{x,0} \cdot e^{-k_{NO_x} \cdot (EGR\lambda^{-1} - EGR_0\lambda_0^{-1})} \cdot C_{NO_x} \cdot C_{dyn} \tag{4.22}$$

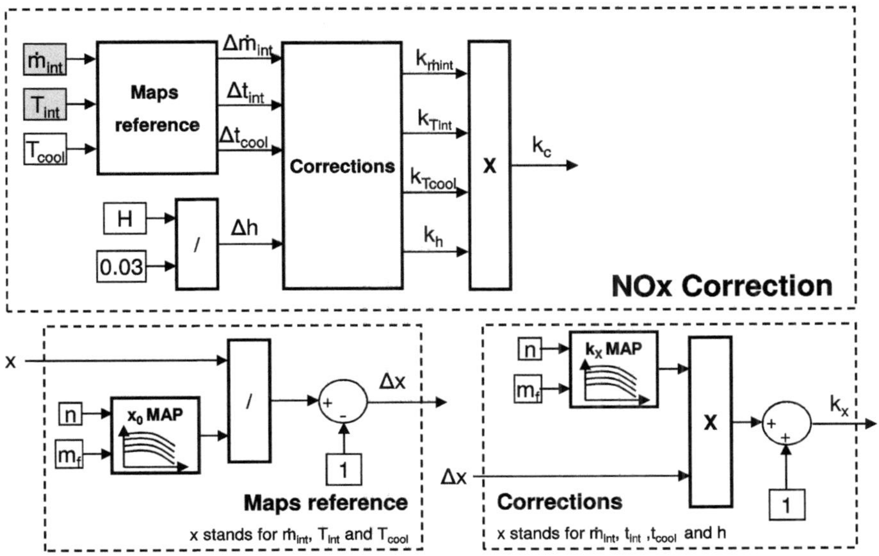

Fig. 4.9 Structure of the factor C_{NO_x} calculation for correcting NO_x output

where C_{NO_x} and C_{dyn} are used as correction factors to the model structure, and are explained in subsequent paragraphs.

4.3.2.3 Correction Factors

The factor C_{NO_x}, defined with the product

$$C_{NO_x} = C_{\dot{m}_{int}} \cdot C_{T_{int}} \cdot C_{T_{cool}} \cdot C_h \tag{4.23}$$

is calculated on the basis of previous experience and tests made in the engine [see (Arrègle et al. 2008) for a sensitivity study of NO_x to relevant engine inputs]. For that, 4 important effects are included in C_{NO_x}: (4.24a) $C_{\dot{m}_{int}}$ indicates variations of the intake mass flow $\dot{m}_{int}$ (Luján et al. 2008); (4.24b) $C_{T_{int}}$ copes with intake temperature T_{int}; (4.24c) $C_{T_{cool}}$ relates engine temperature [especially for distinguishing cold and warm conditions (Broatch et al. 2008); (4.24d) C_h considers Humidity H (if a humidity sensor is available]. The expected NO_x sensitivity for the different parameters is depicted at the right side of (4.24a), (4.24b), (4.24c), (4.24d). Figure 4.9 shows the blocks diagram of the C_{NO_x} model. $k_{\dot{m}_{int}}$, $k_{T_{int}}$, $k_{T_{cool}}$, k_h are calibrated by $[n, m_f]$ dependent maps. If engine is operating at nominal conditions then $C_{\dot{m}_{int}}, C_{T_{int}}, C_{T_{cool}} = 1$. Equations are shown in the following.[1, 2]

$$C_{\dot{m}_{int}} = 1 + k_{\dot{m}_{int}} \left(\dot{m}_{int}/\dot{m}_{int,0} - 1\right) \qquad \uparrow \dot{m}_{int} \downarrow NO_x\ ^{1} \tag{4.24a}$$
$$C_{T_{int}} = 1 + k_{T_{int}} \left(T_{int}/T_{int,0} - 1\right) \qquad \uparrow T_{int} \uparrow NO_x \tag{4.24b}$$
$$C_{T_{cool}} = 1 + k_{T_{cool}} \left(T_{cool}/T_{cool,0} - 1\right) \qquad \uparrow T_{cool} \uparrow NO_x \tag{4.24c}$$
$$C_h = 1 + k_h H/0.03 \qquad \uparrow H \downarrow NO_x\ ^{2}. \tag{4.24d}$$

4.3.2.4 Thermal Loading Model

T_{cool} represents the engine coolant temperature and is measured on-board. T_{cool} is used on-board for defining warm and cold strategies. However, neither the engine block temperature nor the in-cylinder temperature (T_{cyl}) are measured on-board (and rarely in test benches) and are factors of great importance for NO_x formation in diesel engines. If the engine is working on steady-state operation and the variables are already stabilised, T_{cyl} might be approximated as a function only of the operating point conditions and T_{cool} as follows

$$T_{cyl} = f(T_{cool}, n, m_f). \tag{4.25}$$

But if the engine is in transient operation, e.g. a tip-in, it is necessary to consider an extra degree-of-freedom since T_{cyl} will be related with the historic of the engine load. Note that T_{cool} itself is not capable of representing actual T_{cyl} and in addition, it is controlled in closed loop for tracking a reference making that

$$T_{cyl} \neq f(T_{cool}, n, m_f) \tag{4.26}$$

An option is installing a temperature sensor at the exhaust but even though this option should be interesting for improving the model accuracy, these sensors are not usually available on-board.

An alternative is considering the heat conductive problem between the cylinder walls and the coolant. The temperature wall affects the in-cylinder temperature and should be considered as a state of the model. If analysing the problem, as the gas temperature variation (ΔT_{cyl}) state defines the heat power released ($\dot{Q}$)

$$\dot{Q} \quad \propto \quad \Delta T_{cyl} \tag{4.27}$$

[1] If p^*_{boost} set-point is increased for a constant air mass flow ($\dot{m}_a$) set-point, then u_{egr} will be opened and more $\dot{m}_{egr}$ will enter the cylinder (consequently $\dot{m}_{int}$ will be higher), producing that NO_x should be lower. This effect might be shifted if u_{egr} is already opened, especially at low load and regime areas, or engine is running out of the EGR area.

[2] Moisture increases the air specific heat c_p, diminishing in-cylinder temperature T_{cyl}.

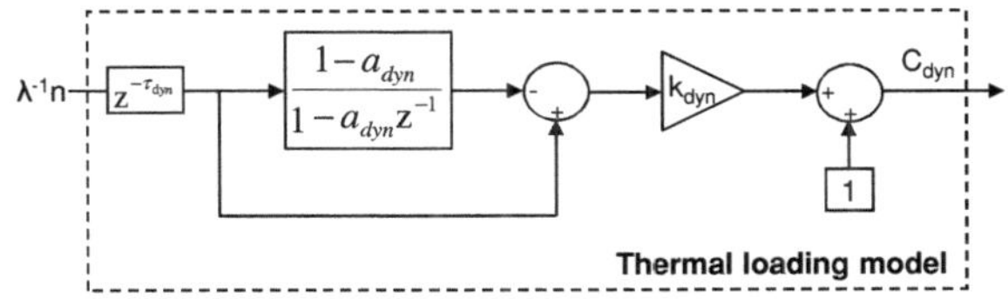

Fig. 4.10 Block diagram of the thermal loading factor C_{dyn}

while the total heat dissipation between the cylinder and the coolant depends on ΔT_{cyl} and the considered time, which is inversely proportional to n, then

$$Q = k\Delta T_{cyl}\frac{1}{n} \tag{4.28a}$$

$$\Delta T_{cyl} \propto \lambda^{-1}n \tag{4.28b}$$

which shows that $\lambda^{-1}n$ is related with the thermal load in the cylinder and could be used as an estimator for the problem.

Therefore, a multiplicative factor C_{dyn} to the NO_x model output is proposed. This factor augments the state vector X and copes with the engine loading and indirectly T_{cyl} effects. The correction factor is designed for correcting the NO_x output only during transient, since steady state effects due to changes in T_{cool} are yet corrected through $C_{T_{cool}}$. The main equation is presented in (4.29) and block diagram shown in Fig. 4.10. C_{dyn} is built with the filtered difference of $\lambda^{-1}n$, which represents engine thermal load variation (other options have also been checked, but this has given the best results for the model, which is in line with the discussion above), and a calibrated gain k_{dyn}. When the engine is steady, so $\lambda^{-1}n$ is, $C_{dyn} = 0$. This factor is set to one for steady-state simulations.

$$C_{dyn} = 1 + k_{dyn}(\lambda^{-1}n)\left(1 - z^{-\tau_{dyn}}\frac{1-a_{dyn}}{1-a_{dyn}z^{-1}}\right). \tag{4.29}$$

Figure 4.11 shows relevant signals during a sharp engine transient in n and m_f. T_{exh} is measured with a sensor and compared with T_{cool} showing two effects: First, T_{cool} is much more filtered and delayed than T_{exh}, and second, the sensitivity of T_{cool} is not sufficient for representing the actual thermal transient occurring on-engine, i.e. T_{cool} hardly varies from 82 to 92 °C while T_{exh} goes from 200 to 700 °C. Bottom plot shows the value C_{dyn} tuned for the NO_x model, being lower than 1 during the transient but finally converging to 1. This is for representing that engine comes from a colder state. In that plot, NO_x sensor (z_{NO_x}) is compared with y_{NO_x} model output [see (4.21)] without C_{dyn} (it overestimates NO_x) and with C_{dyn}.

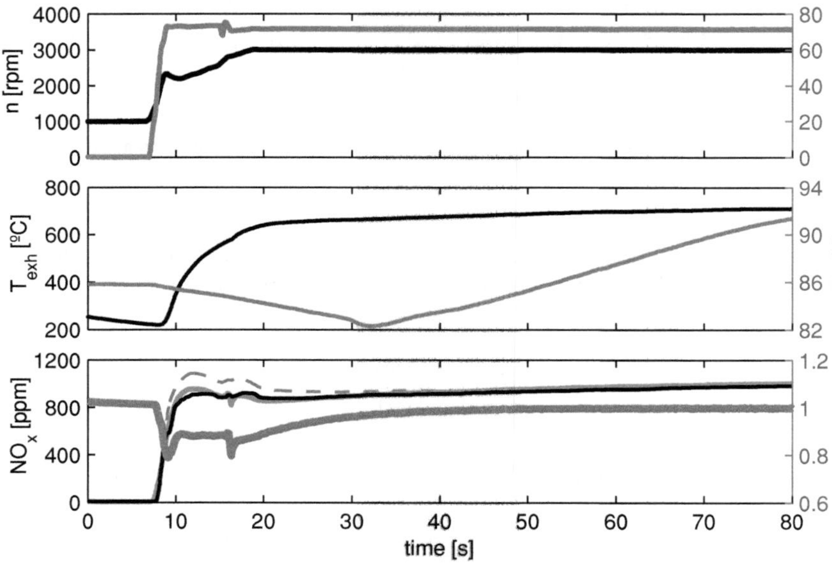

Fig. 4.11 NO_x and thermal response during a sharp transient. *Top plot* n and m_f evolution. *Middle plot* exhaust temperature (T_{exh}) and coolant temperature (T_{cool}) during the transient. Dynamics of both signals, especially for T_{cool}, are slow, which affects final NO_x. *Bottom plot* and *left axis*: *black line* z_{NO_x} by sensor, *gray dashed line* y_{NO_x} with $C_{dyn} = 1$, *thin gray line* y_{NO_x} using C_{dyn}. *Bottom plot* and *right axis*: *thick gray line* C_{dyn} factor calculated by the thermal model

4.3.2.5 The Effect of the Injection Parameters

The injection duration (t_{id}), the start of injection (SOI) and the common rail pressure (p_{rail}) are critical for NO_x emissions (Arsie et al. 2012) and so is the engine speed (n) that is coupled with the others, which here is a model input. For the current structure, the engine commercial calibration for the injection settings is used being the injected fuel mass (m_f) the representative variable. For that reason, all nominal conditions (marked with subscript 0) are related to the set [n m_f], i.e every pair [n m_f] defines nominal $u_{soi,0}$, $u_{prail,0}$ and $u_{id,0}$. The engine torque (M_e) could be used instead of m_f, but even though both M_e and m_f are estimated by the ECU, m_f is preferred as used in the ECU of the considered engine. SOI or p_{rail} could be considered as exponential corrections if different from the nominal, and such effects should be taken into account in the combustion model (this could be useful for controlling the fuel path).

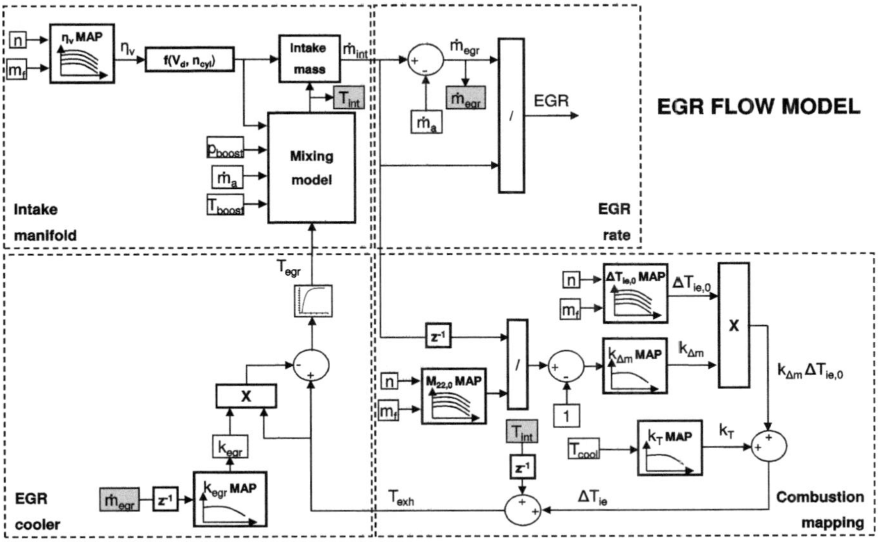

Fig. 4.12 Structure of the EGR flow model

4.3.2.6 The EGR Flow Model

The EGR rate

$$EGR = \frac{\dot{m}_{egr}}{\dot{m}_{int}} \tag{4.30}$$

is calculated on the basis of a mean value model of the volumetric efficiency (η_v) using boost pressure (p_{boost}) and air mass flow ($\dot{m}_a$) inputs coming from commercial sensors. The model block diagram is shown in Fig. 4.12.

The intake manifold mass flow ($\dot{m}_{int}$) is obtained from η_v (mapped with the steady-state tests). p_{boost} is assumed constant in the manifolds that confluence at the intake ($egr,boost,int$) while T_{boost} is measured and T_{exh} calculated by means of a combustion model based on maps and corrections over T_{cool} and $\dot{m}_{int}$.

$$\dot{m}_{int} = \frac{\eta_v \, n_{cyl} \, V_d \, n \, p_{boost}}{2 \, R \, T_{int}} \tag{4.31}$$

n, m_f, T_{boost} and $\dot{m}_a$ are measured by sensors, the unit displaced volume (V_d) and the number of cylinders (n_{cyl}) are known beforehand and a model is proposed for T_{int}. For instance, Wahlström and Eriksson (2011) propose an isothermal model ($T_{int} = T_{boost}$); for the current work a mixing model based on the energy conservation at the intake junction is utilised, i.e. intake volume is small and mixing is so fast that no heat is transferred to the walls.

$$\dot{m}_{int}T_{int} = \dot{m}_a T_{boost} + \dot{m}_{egr}T_{egr} \tag{4.32}$$

and from (4.31), the product $\dot{m}_{int}T_{int}$, defining the intake enthalpy flow, can be inferred as

$$\dot{m}_{int}T_{int} = \frac{\eta_v\, n_{cyl}\, V_d\, n\, p_{boost}}{2R} \tag{4.33}$$

and rewriting the mass equilibrium at the intake junction

$$\dot{m}_{egr} = \frac{\eta_v\, n_{cyl}\, V_d\, n\, p_{boost}}{2RT_{int}} - \dot{m}_a \tag{4.34}$$

and finally substituting in (4.32), T_{int} can be solved

$$T_{int} = \frac{\eta_v\, n_{cyl}\, V_d\, n\, p_{boost}\, T_{egr}}{\eta_v\, n_{cyl}\, V_d\, n\, p_{boost} + 2\, R\, m_{boost}\left(T_{egr} - T_{boost}\right)}. \tag{4.35}$$

Two additional models (see Fig. 4.12) are needed in order to calculate T_{egr} (not usually measured on commercial engines): a T_{exh} model and an EGR cooler efficiency model.

4.3.2.7 T_{exh} Model

Temperature increase between the intake and the exhaust ΔT_{ie} is calculated from a steady map and corrections

$$T_{exh} = T_{int}z^{-1} + \Delta T_{ie} \tag{4.36}$$

adding the discrete delay z^{-1} for coping with causality.

ΔT_{ie} can be calculated on the basis of heat release functions (e.g. Wiebe for Watson functions) and solving the principle equations of the cylinder, or fitted to experimental values. Since the results of such physical approach may be calculated beforehand, a map based model coherent with the full RT NO_x model is proposed with two corrections based on $\dot{m}_{int}$ and T_{cool}

$$\Delta T_{ie} = \Delta T_{ie,0} \cdot k_{\Delta m} + k_T \tag{4.37}$$

Maps depending on $[n, m_f]$ are used for generating references for $\Delta T_{ie,0}$ and $\dot{m}_{int,0}$. Then Δm is calculated

$$\Delta m = \frac{\dot{m}_{int}z^{-1}}{\dot{m}_{int,0}} - 1 \tag{4.38}$$

and defining the calibrated map output

$$k_{\Delta m} = f(\Delta m) \tag{4.39}$$

Finally, term k_T represents the steady-state effect of the wall cylinder temperature effect and is calculated from the curve

$$k_T = f(T_{cool})$$

noting that T_{cool} adds a degree of freedom to the problem as there exists a closed loop control that manages the coolant mass flow and the set-point for T_{cool}. The difference with respect to the thermal factor C_{dyn} is that the latter copes with transients, while k_T with the static effect.

If SOI and p_{rail} are different from the nominals used for generating the nominal maps, then the additional factors k_{SOI} and $k_{p_{rail}}$ should be calibrated in the combustion model.

4.3.2.8 EGR Cooler Model

T_{egr} is calculated by a temperature drop and a first order discrete filter with response time a_{egr} and a transport delay τ_{egr}

$$T_{egr} = z^{-\tau_{egr}} \frac{1 - a_{egr}}{1 - a_{egr} z^{-1}} \left(T_{exh} - k_{egr} \left(T_{exh} - T_{cool}\right)\right) \tag{4.40}$$

where k_{egr} is calculated by the curve

$$k_{egr} = f(\dot{m}_{egr}) \tag{4.41}$$

and τ_{egr} is function of the engine characteristic time

$$\tau_{egr} = \frac{k_{\tau_{egr}}}{n}. \tag{4.42}$$

4.3.3 Tuning Methodology

The tuning and validation methodology is shown in Fig. 4.13, splitting the steady-state and dynamic testing. Dashed lines show critical paths, e.g. once that NO_x structure has been fixed after preliminary studies that include $[CO_2]_{int}$ and sensitivity analysis, changes in the NO_x structure are expected to be minimum.

Fig. 4.13 Scheme of the calibration and validation procedure for the NO_x model

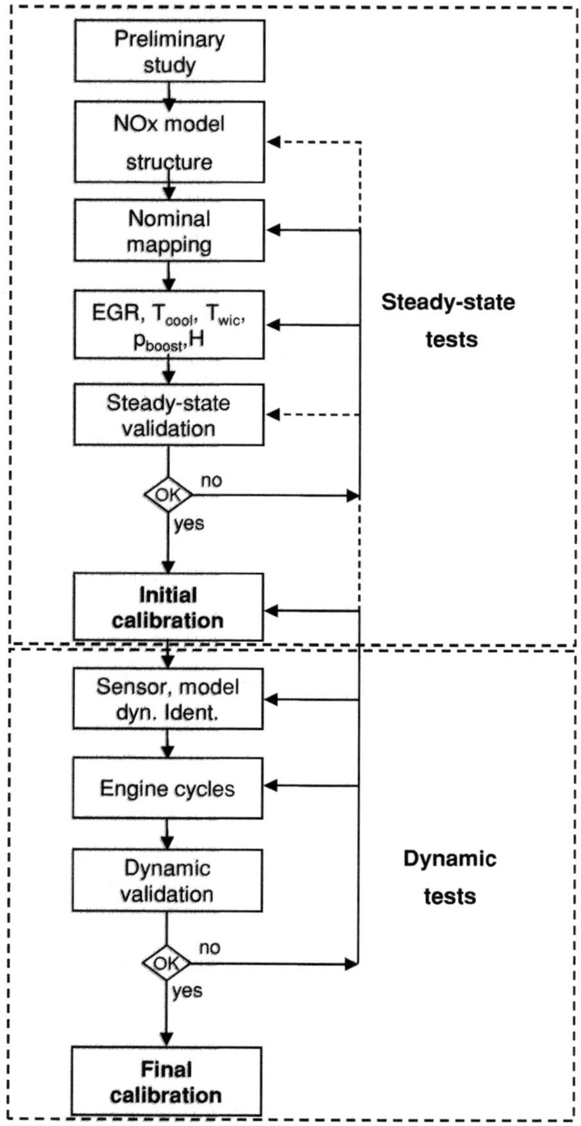

4.3.3.1 Static Tuning

The steady tests are considered for having a representative sample of the operative engine range and calibrating nominal maps, main parameters and curves by using LS, with the following sequence:

Table 4.1 Mean errors when fitting the model with different input variables

	$EGR\lambda^{-1}$	Corrected Ip	$[CO_2]_{int}$ from gas analyser
g/h	9.31	9.78	8.54
%	8.09	8.96	7.58

$[CO_2]_{int}$ from the gas analyser is the mean of the 4 extractions of the intake manifolds as discussed in Sect. 3.2.1. $EGR\lambda^{-1}$ is calculated on the basis of the EGR mean value flow model and the λ^{-1} model. Signal from the intake lambda probe (I_p) is corrected by pressure and some data points with faulty measurements have been removed because they present problems with the sensor saturation. NO_x output is adjusted in g/h

- From all tests, η_v, $\dot{m}_{egr}$ and temperatures differences are calculated and used for fitting the maps as function of the engine speed and fuel mass.
- From nominal tests, maps are built for $NO_{x,0}$, $EGR_0\lambda_0^{-1}$ and $T_{cool,0}$ as function of the engine speed and fuel mass.
- From all tests, the RT NO_x model is simulated in order to estimate the results for $\dot{m}_{int}$ and T_{int}. The results are used for identifying $\dot{m}_{int,0}$ and $T_{int,0}$.
- The ratios $\dot{m}_{int}/\dot{m}_{int,0}$, $T_{int}/T_{int,0}$ and $T_{cool}/T_{cool,0}$ are calculated from all tests.
- Curves k_{NO_x}, k_m, $k_{T_{int}}$, $k_{T_{cool}}$ and k_h are fitted by LS of the main model equation (4.22). This adjustment is made independently by using the steady-state data set. In addition to $EGR\lambda^{-1}$ and in order to compare results with other possibilities for estimating $[O_2]_{int}$, I_p from an intake UEGO sensor and $[CO_2]_{int}$ from the gas analyser are evaluated. The results are presented in Table 4.1. The best fit is for $[CO_2]_{int}$ but this is not available on-board, while it can be used for tuning and validation. Therefore, the minimum error is achieved with the variable $EGR\lambda^{-1}$, calculated in (4.14).

4.3.3.2 Dynamic Tuning

Dynamic tests are used for adjusting the C_{dyn} model, the path dynamics (e.g. EGR path) and for the fine tuning of the maps and curves fitted with the steady-state tests. Furthermore, the step tests are used for calibrating the sensors; and the cycles for validating the model in dynamic state. If the validation is not satisfactory, the dynamic cycles can be repeated and identifications checked. If the error persists, it could be advisable to come back to the static calibration and re-tune the unreliable parameters (see Fig. 4.13).

Even though not included in the scheme, tests from different campaigns could lead to drift because of the ageing or other effects that are not well modelled. Anyway, the repetition of a specific DOE will not avoid the model drift, and for sure, it is not a cost-effective solution. For that, adaptive strategies are useful as discussed in next chapters.

4.3.4 NO_x Model Results

In the following, NO_x model results with static and dynamic tests are presented. Furthermore, error metrics are built in order to determine the model accuracy.

4.3.4.1 Steady-State Results

Top plots of Fig. 4.14 show x_{NO_x} and z_{NO_x} (from the NO_x-ZrO_2 sensor) for both ppm and g/h units as well as histogram plots of relative and absolute errors. The mean error is 9.31 g/h, with a minimum deviation of σ=14.74 g/h proving the validity of the calibration. The results by evaluating the ppm output present a mean error of 30.43 ppm and σ = 54.84 ppm. It must be noted that errors of 1 % in input variables can produce final NO_x errors of 33 %, measured in a predictive and physics-based NO_x model [see the NO_x sensitivity study published in Arrègle et al. (2008)]. The maximum error in a DDM, such as the one presented here, is bounded by the tables grid and only the ageing and dispersion errors can affect, problems that exist also in more complex models. This fact and the results presented in this subsection justify the validity of a data-driven NO_x model, as well as the efficient computation and simple programming.

Medium and bottom plots of Fig. 4.14 show absolute and relative error metrics for both outputs in g/h and ppm. Again, only a minimum percentage of points (total are 363) are out of the $\pm\sigma$ boundary, also plotted in those plots. If a higher accuracy on ppm output is required, tuning objective can be easily shifted to ppm.

Additional information from the model can be extracted to compare intermediate variables (calculated by the model) and the corresponding measurements, as for instance for the mean value EGR flow model. A multi-objective optimisation could be programmed if other model states should be important for control. However, here x_{NO_x} output error has been minimised. The left plot of Fig. 4.15 compares *EGR* calculated by the model and by using $[CO_2]_{int}$ averaged from the 4 extractions of the in-cylinder manifolds. The fitting is acceptable in the whole range. Results for T_{egr} are shown in the right plot of Fig. 4.15, and the fitting in general is also acceptable, but there exists a high dispersion around the centre line. This is mainly due to areas with low or null EGR, where T_{egr} does not make any sense ($\dot{m}_{egr}$=0).

4.3.4.2 Model Robustness

This section turns out the attention to the model robustness analysing the model sensitivity to errors in the main parameters and inputs. 2 % disturbances in the parameters presented in Table 4.2 are applied and the final variation in NO_x model output is stored, e.g. a change of 2 % in n produce an error of 4.51 % in NO_x. Parameters with more influence on model output are n, m_f and $\dot{m}_a$. These signals basically represent the engine operating conditions, and define the EGR setting (jointly with p_{boost} that

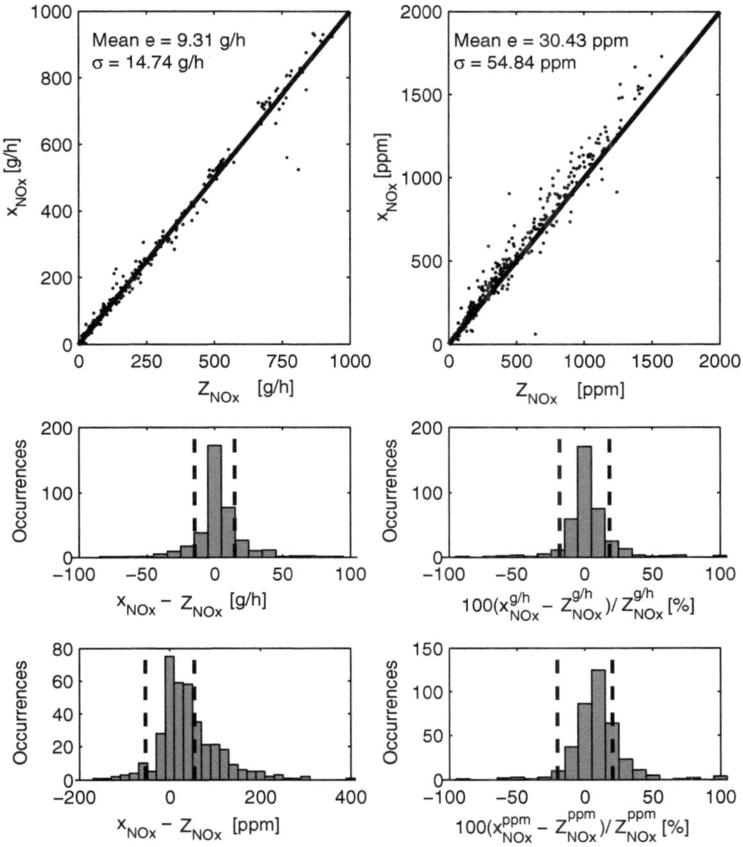

Fig. 4.14 RT NO_x model results with steady tests. *Top plots* Model fitting with respect to the NO_x sensor output. *Medium plots* Absolute and relative error in percentage with respect to the NO_x sensor output and for g/h output ($x^{g/h}_{NO_x}$) of the model. *Bottom plots* Absolute and relative error in percentage with respect to the NO_x sensor output and for ppm output ($x^{ppm}_{NO_x}$) of the model

presents a minor influence) and the trend was already expected. p_{boost} and η_v have the same influence as expected because both are multiplied for calculating $\dot{m}_{int}$ and consequently $\dot{m}_{egr}$.

An important conclusion of this study is that variables with more influence on NO_x are directly measured, and then having a reliable measurement is crucial for the model accuracy. In the second group, $NO_{x,0}$, $\dot{m}_{int}$ and η_v steady maps denote the most relevant maps to calibrate. The rest of variables have lower influence but not negligible.

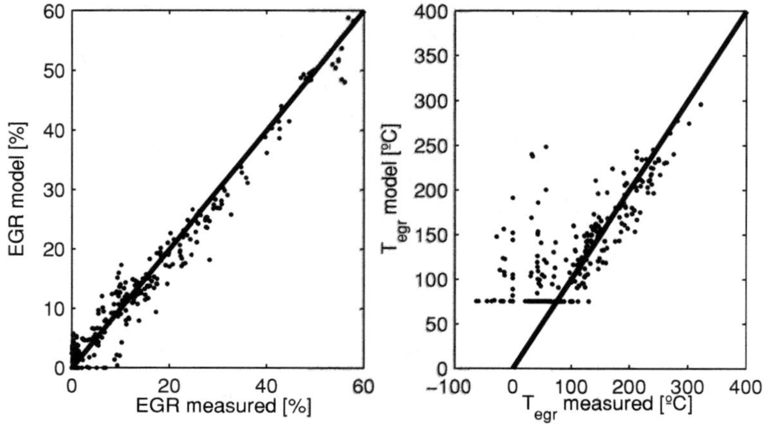

Fig. 4.15 Relevant results from the EGR flow model. *Left plot* EGR calculated by the model and from the gas analyser measurements. *Right plot* T_{egr} calculated by the model and measured by a sensor

Table 4.2 Results over the final NO_x model output of applying variations of 2 % on different variables

Variable	Variation (%)	Mean error (%)	STD error (%)	Max. error (g/h)	Min. error (g/h)
n	+2	4.51	3.27	15.23	−1.41
m_f	+2	3.67	2.53	13.42	−0.36
m_{boost}	+2	3.42	3.50	12.60	−1.12
$NO_{x,0}$ MAP	+2	2	0.00	2.00	2.00
$\dot{m}_{int}$ MAP	−2	1.79	0.74	3.16	0.06
η_v MAP	−2	1.03	2.78	8.29	−2.93
p_{boost}	−2	1.03	2.78	8.29	−2.93
T_{cool}	+2	0.97	0.39	1.81	0.08
$EGR_0\lambda_0^{-1}$	+2	0.66	0.79	3.22	0.00
k_{egr} MAP	−2	0.16	0.31	1.39	−0.51
ΔT_{ie} MAP	+2	0.06	0.14	0.97	−0.14
k_{NO_x} MAP	+2	0.04	0.48	2.46	−2.42

4.3.4.3 Dynamic Results

Figure 4.11 already showed the ability of the NO_x model to fit the sensor output and the necessity of utilising the factor C_{dyn}. Figure 4.16 shows the model results with the SDMP test. This test is a good benchmark for testing the dynamic ability of the model as fast transients are performed. The influence of C_{dyn} is remarkable and can be seen in the left plot of the figure. Thin gray line corresponds to the model output but cancelling the effect of the thermal dynamic factor, i.e. $C_{dyn} = 1$, whilst the thick gray line is y_{NO_x}. If using $C_{dyn} = 1$, NO_x is overestimated (about 20 % higher than expected) due to the model does not consider the rapid load variations and the magnitude of them. In the SDMP, the influence of C_{dyn} is of crucial importance as

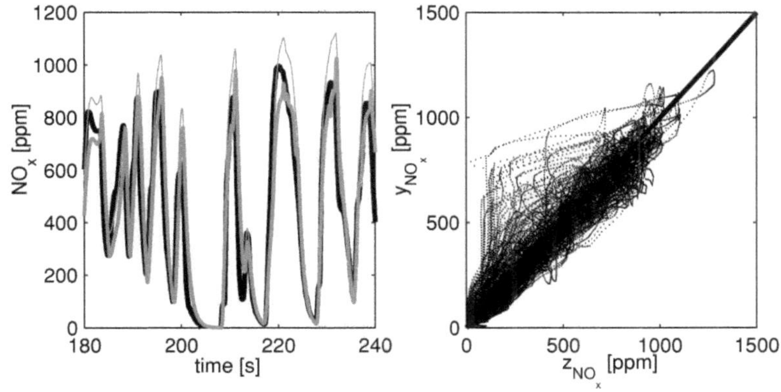

Fig. 4.16 *Left* y_{NO_x} calculated normally (*thick gray*) and cancelling thermal dynamic factor; i.e. $C_{dyn} = 1$ (*thin gray*), and z_{NO_x} (*black*) during a part of the SDMP B5. *Right* y_{NO_x} against z_{NO_x}

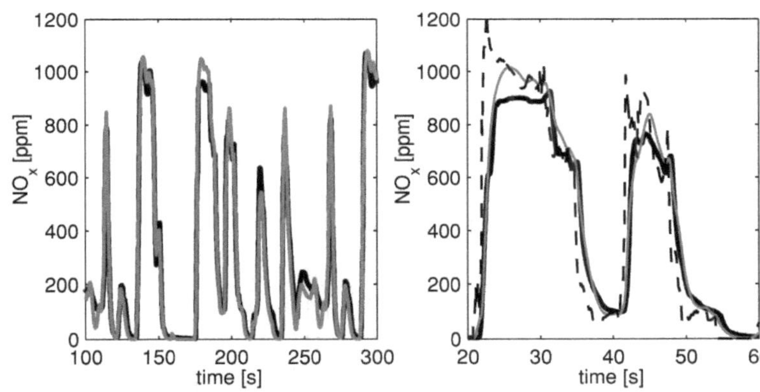

Fig. 4.17 *Left* y_{NO_x} (*gray*) and z_{NO_x} (*black*) in the TRAN A ($T_{cool} = 75\,°C, \Delta u_{egr} = 5\,\%, \Delta p_{boost} = 0$). *Right* x_{NO_x} (*dashed black*), y_{NO_x} (*thin gray*) and z_{NO_x} (*black*) in the same test. x_{NO_x} is the actual NO_x, no delay nor filtering are expected

the profile is ranging from low to high loads (thermal transients are critical). The right plot compares y_{NO_x} with z_{NO_x} obtaining a good fitting indeed when using ppm as output. The mean error in the SDMP tests (B5 and B6) is in the order of 50 ppm, which can be minimised by using online observers.

Figure 4.17 shows results on the TRAN A ($T_{cool} = 75\,°C$, $\Delta u_{egr} = 5\,\%$, $\Delta p_{boost} = 0$) cycle; the left plot compares y_{NO_x} with z_{NO_x} while the right plot includes the x_{NO_x} estimation. The sensor model fits well enough to have a reliable x_{NO_x}, that is observed by cancelling sensor filtering and lag. Even though the NO_x model is based on quasi-static maps, system and physics dynamics are contained in the model inputs, which are directly measured from sensors. As shown in this Figure, the results are also promising.

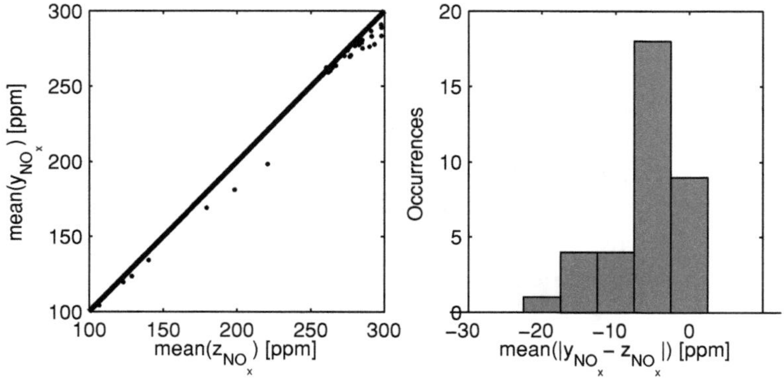

Fig. 4.18 *Left* mean value of NO_x emissions, comparing the sensor z_{NO_x} against the model y_{NO_x} by simulating the NO_x model in all the TRAN cycles. *Right* absolute error bar distribution of the model against the sensor mean($|y_{NO_x} - z_{NO_x}|$). The mean absolute error is less than 10 %

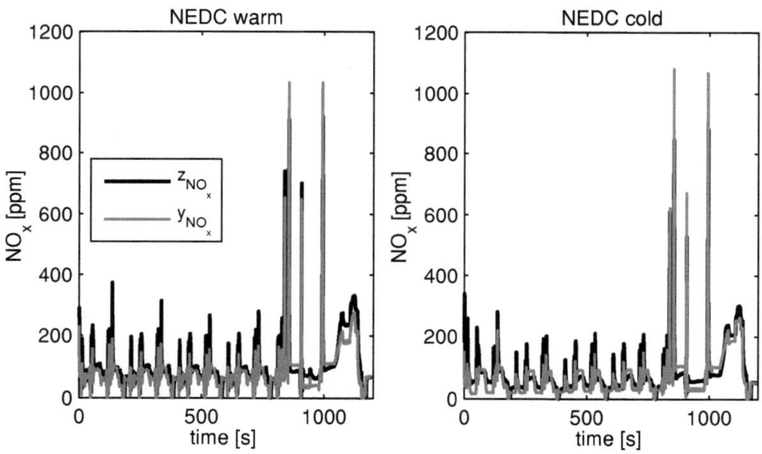

Fig. 4.19 y_{NO_x} model output and z_{NO_x} measured by the sensor in the NEDC. *Left* NEDC warm cycle A1. *Right* cold cycle A2

The absolute mean value for z_{NO_x} and y_{NO_x} is calculated for every FTP cycle

$$mean(y_{NO_x}) = \frac{1}{l_{tran}} \sum_{k=1}^{l_{tran}} (abs(y_{NO_x}(k))) \tag{4.43a}$$

$$mean(z_{NO_x}) = \frac{1}{l_{tran}} \sum_{k=1}^{l_{tran}} (abs(z_{NO_x}(k))) \tag{4.43b}$$

where l_{tran} is the total number of samples for every TRAN cycle. Results are plotted in Fig. 4.18, where the left plot shows $mean(y_{NO_x})$ against $mean(z_{NO_x})$ while the right plot shows the error distributions in ppm. The results demonstrate the model accuracy.

Finally, Fig. 4.19 shows the results with the NEDC A1 (warm) and A2 (cold). Especially during the highway part, there exists a bias in both cycles. Since the model is optimised for a set of steady-state tests, dispersion effects or errors due to non-modelled variables (NO_x is highly affected by the ambient conditions) can affect the model accuracy. Admittedly, the model dynamics are able to reproduce z_{NO_x} and then the drift can be corrected by using observers as discussed in subsequent chapters.

4.4 Conclusions

An ECU oriented NO_x model has been developed by considering that variations of $[O_2]_{int}$ cause exponential changes in the nominal NO_x, which is conveniently mapped. A study on different alternatives for estimating $[O_2]_{int}$ is presented in the chapter, selecting the inert gas rate $EGR\lambda^{-1}$ as the most convenient. A set of corrections and factors with respect to nominal operating points are included for considering the most sensitive parameters. Furthermore, a thermal dynamic model for simulating thermal loading is proposed as function of the product $\lambda^{-1}n$. Two additional models are needed for the NO_x model: an EGR gray box model for estimating EGR and a model for estimating the fuel-to-air ratio λ^{-1}.

A total of 363 steady-state tests are performed for calibrating all parameters and maps. Step-like tests for system and sensors identifications are made and different cycles are tested in order to validate the model. The NO_x prediction is successful and the simple programming as well as the fact that all required signals are available in commercial ECUs makes the model suitable for online engine control or diagnosis.

References

Andersson M (2006) Fast NO_x prediction in diesel engines. PhD thesis, Lund University

Andersson M, Hultqvist A, Johansson B, Nöhre C (2006) Fast physical NO_x prediction in diesel engines. In: The diesel engine: the low CO_2 and emissions reduction challenge (conference proceedings), Lyon

Arrègle, J, López JJ, Guardiola C, Monin C (2008) Sensitivity study of a NO_x estimation model for on-board applications. SAE tecnhical report 2008-01-0640

Arrègle J, López JJ, Guardiola C, Monin C (2010) On board NO_x prediction in diesel engines: a physical approach. In: del Re L et al (eds) Automotive model predictive control: models, methods and applications. Springer, Berlin. ISBN-1849960704

Arsie I, Di Leo R, Pianese C, De Cesare M (2012) Combustion noise and pollutants prediction for injection pattern and EGR tuning in an automotive common-rail diesel engine. In: ECOSM 2012 workshop on engine and powertrain control, simulation and modeling

Broatch A, Luján JM, Serrano JR, Pla B (2008) A procedure to reduce pollutant gases from diesel combustion during European MVEG-A cycle by using electrical intake air-heaters. Fuel 87(12):2760–2778

Deng J, Stobart R, Liu C, Winward E (2012) Explicit model predictive control of the diesel engine fuel path. SAE technical paper 2012-01-0893

Desantes JM, Galindo J, Guardiola C, Dolz V (2010) Air mass flow estimation in turbocharged diesel engines from in-cylinder pressure measurement. Exp Thermal Fluid Sci 34(1):37–47

Finesso R, Spessa E (2012) Real-time predictive modeling of combustion and NO_x formation in diesel engines under transient conditions. SAE technical paper 2012-01-0899

Galindo J, Climent H, Guardiola C, Doménech J (2009) Strategies for improving the mode transition in a sequential parallel turbocharged automotive diesel engine. Int J Autom Technol 10(2):141–149

Guardiola C, Pla B, Blanco-Rodriguez D, Mazer A, Hayat O (2013) A bias correction method for fast fuel-to-air ratio estimation in diesel engines. Proc Inst Mech Eng Part D J Automob Eng 227(8):1099–1111

Lee J, Lee S, Park W, Min K, Song HH, Choi H, Yu J, Cho SH (2013) The development of real-time NO_x estimation model and its application. SAE technical paper 2013-04-08

Luján JM, Galindo J, Serrano JR, Pla B (2008) A methodology to identify the intake charge cylinder-to-cylinder distribution in turbocharged direct injection diesel engines. Measur Sci Technol 19(6):065401

Macián V, Luján JM, Guardiola C, Perles A (2006a) A comparison of different methods for fuel delivery unevenness detection in diesel engines. Mech Syst Sig Process 20(8):2219–2231

Macián V, Luján JM, Guardiola C, Yuste P (2006b) DFT-based controller for fuel injection unevenness correction in turbocharged diesel engines. IEEE Trans Control Syst Technol 14(5):819–827

Payri F, Guardiola C, Blanco-Rodriguez D, Mazer A, Cornette A (2012a) Methodology for design and calibration of a drift compensation method for fuel-to-air tatio estimation. SAE technical paper 2012-01-0717

Payri F, Guardiola C, Pla B, Blanco-Rodriguez D, Mazer A, Cornette A (2012b) Fast estimation of fuel-to-air ratio through the combination of a fast model and an exhaust lambda sensor. In: SIA 2012 international conference

Payri F, Luján JM, Guardiola C, Rizzoni G (2006) Injection diagnosis through common-rail pressure measurement. Proc Inst Mech Eng Part D J Automob Eng 220(3):347–357

Wahlström J, Eriksson L (2011) Modelling diesel engines with a variable-geometry turbocharger and exhaust gas recirculation by optimization of model parameters for capturing non-linear system dynamics. Proc Inst Mech Eng Part D J Automob Eng 225(7):960–986

Part III
Adaptive Filtering for Observing NO_x and λ^{-1}

Chapter 5
Adaptive Observers for the Dynamic Estimation of Engine Variables

5.1 Introduction

In the previous chapters, different sensors and models have been explored in order to estimate engine variables, with focus on the exhaust λ and NO_x. From all the possible problems in the signals, the dynamic filtering, delay and drift are noticeable from sensor and model outputs (Guardiola et al. 2011, 2013b, c). These issues are affected not only by sensors and models themselves, but by the engine layout.

5.1.1 Sensor Layout

Figure 5.1 shows three possible layouts for on-board gas concentration sensors in order to estimate the engine-out NO_x and λ^{-1} for OBD and control of the AT devices. The final implantation depends heavily on the AT configuration and a variety of solutions were already discussed in the Chap. 2. Manufacturers are struggling for finding the configuration of systems and sensors that optimises the trade-off between low emissions, efficient torque production, low cost and robustness.

The usual position of UEGO sensors is directly downstream of the turbine: this way the sensor is near of the source and avoids the large range of pressure variation that exists in the exhaust manifold. However, if a SCR is installed, a NO_x sensor is needed for their control. Such sensor might be placed downstream of the AT devices in order to measure tail-pipe emissions, as sketched in the options 1, 2 and 3 of Fig. 5.1. Hence it seems possible to avoid the UEGO sensor and just rely on the NO_x sensor information, which measures λ^{-1} and NO_x, as depicted in the third exhaust line of Fig. 5.1. However, the slow response of the ZrO_2 sensor, if compared with that of a UEGO sensor, is poor for RT purposes as was shown in Fig. 2.5.

D. Blanco-Rodriguez, *Modelling and Observation of Exhaust Gas Concentrations for Diesel Engine Control*, Springer Theses, DOI: 10.1007/978-3-319-06737-7_5,

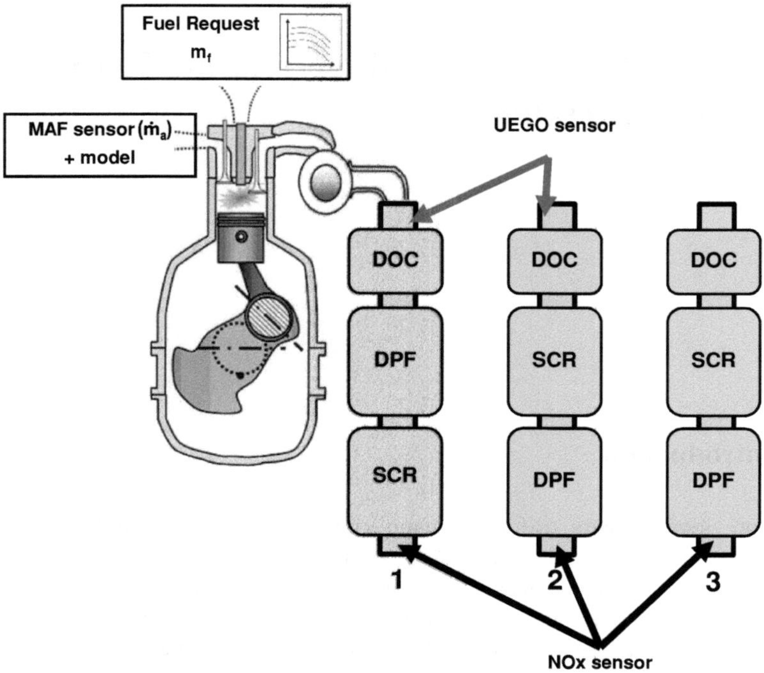

Fig. 5.1 Possible UEGO and NO_x sensors configuration setup for a turbocharged diesel engine

5.1.2 The Drift Problem on Models

Models can be used instead for estimating variables but despite of the promising results with the models presented in Chap. 4, two problems can be underlined. On one hand, the model accuracy is driven by the collection of the appropriate data and calibration of all the parameters. This is a hard and time consuming task. In fact, the ECU has a big number of maps and parameters for the engine and vehicle management. On the other hand and independently of how well the model has been calibrated, there is inevitably a drift between the system and the model as the surrounding conditions change and the engine ages (this occurs with all kind of models). DDMs are highly sensitive to the calibration data set and will have problems with ageing, manufacturing discrepancies, slowly varying parameters and other non-modelled variables.

Figure 4.19 showed the NO_x model (filtered) output y_{NO_x} for two NEDC cycles (warm A1 and cold A2) and a clear drift exists, especially during the highway part in the A1 and during the whole cycle in A2. Another example is shown in Fig. 5.2, where an old and non-optimised calibration set for the model (maps and sensor dynamics) is used for simulating the SDMP B6 test. The dynamics are well caught but a clear drift also exists. In this case it is not only ageing that affects, but test or ambient conditions.

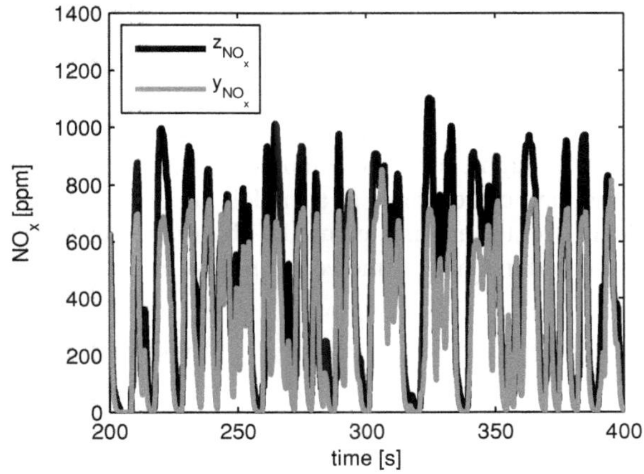

Fig. 5.2 Model output y_{NO_x} and NO_x sensor signal z_{NO_x} in the SDMP B6 test with a non-optimized calibration data set. Even though the dynamics are well caught, y_{NO_x} is biased with respect to z_{NO_x}

5.1.3 The Use of Adaptive Strategies

The dynamic problems of the sensors and the drift on models motivate the design of adaptive on-board strategies. Concretely, two kind of models and algorithms are proposed in order to solve these issues, which are:

- augmented models for drift correction and dynamic estimation of variables, and
- learning algorithms for updating look-up tables and parameters on the basis of COMs, which can be used for calibration and/or online adaptation.

In the next, the dissertation focuses on the description of the proposed methods, while Chap. 6 will give the results with experimental data.

5.2 Augmented Models for Drift Correction

The design of augmented models for bias tracking and the implementation of a KF for observing this bias is discussed. A known case is considering a fast model that keeps high frequency components of the considered variable while a slow but steady-state accurate sensor permits to correct the model drift. This problem is solved by observing the bias θ between a reference measurement z and an initial estimation u, considering the measurement dynamics and delay (Guardiola et al. 2013c; Payri et al. 2012).

5.2.1 Drift Correction Algorithm

The model is presented in the discrete state-space form. The sample time is considered constant and the model equation is defined as follows

$$x_k = \mathbf{F}x_{k-1} + Bu_k + \mathbf{W_k} \tag{5.1a}$$

$$z_k = Hx_k + v_k \tag{5.1b}$$

Without loss of generality, the problem can be stated for estimating the drift $\theta_k \in \mathbb{R}$, augmenting the standard model with such state. $u_k \in \mathbb{R}$ is the state-space input, and corresponds to a model output with a certain delay τ. The effect of the delay over the augmented model observation is discussed in Chap. 6. For the models presented in Chap. 4, the vector u is as follows

$$u_k^{NO_x} = x_{NO_x}(k - \tau_{NO_x}/T_s) \tag{5.2a}$$

$$u_k^{\lambda^{-1}} = x_\lambda^{-1}(k - \tau_{\lambda^{-1}}/T_s) \tag{5.2b}$$

depending if NO_x or λ^{-1} is the key variable. The sensor signal $z_k \in \mathbb{R}$ the output, while the first order filter presented in Eq. (3.3) with response time a is used for modelling z_k, giving the signal $x_{fk} \in \mathbb{R}$. The state vector $x_k \in \mathbb{R}^2$ is built as follows

$$x_k = \begin{bmatrix}\theta_k & x_{fk}\end{bmatrix}^T \tag{5.3}$$

$\mathbf{F}$, B and H are the constant state-space matrices

$$\mathbf{F} = \begin{bmatrix} 1 & 0 \\ 1-a & a \end{bmatrix} \quad B = \begin{bmatrix} 0 \\ 1-a \end{bmatrix} \quad H = \begin{bmatrix} 0 & 1 \end{bmatrix} \tag{5.4}$$

Noises $W_k \in \mathbb{R}^2$ with $w_k \in \mathbb{R}$

$$W_k = [w_k \quad 0]^T \tag{5.5}$$

and $v_k \in \mathbb{R}$, from process and output respectively, correspond to Gaussian processes with the following stationary constant covariance matrices

$$\mathbf{Q} = cov\left[ww^T\right] = \begin{bmatrix} \sigma_w^2 & 0 \\ 0 & 0 \end{bmatrix} \tag{5.6a}$$

$$r = cov\left[vv^T\right] = \sigma_v^2 \tag{5.6b}$$

The system is fully observable and linear time invariant (LTI). Note that only a noise to the state θ is applied as far as this is the key observed variable and the noise is directly transmitted to x_f. Applying noise to x_f adds an extra degree of freedom that does not necessarily improve the identification. The system can easily be augmented with other key variables if necessary. In such sense, the state vector X (4.18) and the input vector U (4.19) from the RT NO_x model could be included in the vectors of the state-space model (5.1a), (5.1b) for a compact programming.

5.2.1.1 The Kalman Filter for Drift Observation

The optimal solution of the estimation problem was addressed by Kalman (1960), and has been applied by different authors to the automotive domain (see for example Alberer and del Re 2009; Desantes et al. 2011; Schilling et al. 2006, 2008). In a first step, the state vector is predicted considering the system input and their expected dynamic characteristics and in a second step, this is updated by calculating the estimation error e_k and weighting it by means of the KF gain K_k

$$\hat{x}_{k|k-1} = \mathbf{F}\hat{x}_{k-1} + Bu_k \tag{5.7a}$$

$$e_k = z_k - H\hat{x}_{k|k-1} \tag{5.7b}$$

$$\hat{x}_k = \hat{x}_{k|k-1} + K_k e_k \tag{5.7c}$$

Kalman gain K_k was solved in the Eq. (2.13).

5.2.1.2 The Steady-State Kalman Filter

Since current application considers a linear time invariant (LTI) and a fully observable system, the filter is steady-state (Kalman and Bucy 1961). A constant Kalman gain K_∞

$$K_\infty = [k_1 \quad k_2]^T \tag{5.8}$$

is then considered and calculated offline

$$K_\infty\left(\sigma_w^2/\sigma_v^2, a\right) = \lim_{k\to+\infty} K_k \tag{5.9}$$

The use of a constant K_∞ leads to a sub-optimal filter neglecting the K updating (Simon 2001) but alleviating the computational burden, since all the iterative process for obtaining K_∞ is performed offline. Anyway, if a more complex sensor model (including the effect of the operating conditions on the sensor dynamics) were employed, the Kalman gain would vary with time, and then K_k might be approximated online by means of the steady-state solution of the Riccati Eq. (2.13):

$$k_1 = \sqrt{(1-k_2)\frac{\sigma_w^2}{\sigma_v^2}} \tag{5.10a}$$

$$k_2 = \frac{s}{s+1} \tag{5.10b}$$

with

$$s = (1-a)\sqrt{\frac{\sigma_w^2/\sigma_v^2}{1-k_2}(1+a-ak_2) + a^2k_2} \tag{5.11}$$

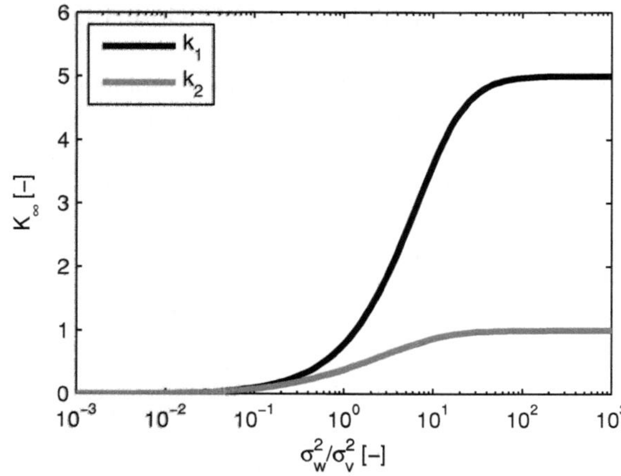

Fig. 5.3 Evolution of the elements of K_∞ gain vector when varying σ_w^2/σ_v^2. Coordinate axis is in logarithmic scale and $a = 0.8$

which is demonstrated in the Appendix A.3.2. k_2 is solved by a fixed point iteration, and k_1 by direct substitution of k_2.

5.2.1.3 Final Estimation

The final estimation $\hat{x}$ on a delayed basis is computed by adding the observed bias to the input u [see (5.2a), (5.2b)]

$$\hat{x}(k - \tau/T_s) = u_k + \hat{\theta}_k \tag{5.12}$$

This equation might be included in the state-space model (see Sect. 6.3.3); the online estimation solving the delay problem is discussed later (see Sects. 6.2.3 and 6.3.3).

5.2.2 Observer Tuning

The KF is tuned with the adequate selection of the quotient σ_w^2/σ_v^2 that affects the model and sensor respectively, defining an optimal K_∞. Figure 5.3 shows the different values of the two elements of K_∞ for different values of σ_w^2/σ_v^2. k_1 is the bias gain, while k_2 corrects the sensor model. Note that the lower the sensor noise is considered (higher σ_w^2/σ_v^2), the faster the states' estimations are updated (because sensor measurement is propagated to the states). The filter behaves as a low pass filter for the sensor and a high pass filter for the model.

Hence the filter tuning defines how fast the bias is cancelled, but two issues must be considered:

- The filter also rejects the sensor noise. Using a very high Kalman gain implies do not filtering the measurement noise.

- If the sensor dynamics characteristics are not perfectly known, a high value of the gain causes the system to rely on incorrect information, and then peaks appear in the estimation especially during transients.

For illustrating the latter issue, the augmented model (5.1a), (5.1b) is simulated by using synthetic signals. A step transition in the objective variable, named $x_r \in \mathbb{R}$ is produced by defining the following parameters for the sensor model

$$a' = a \pm \sigma_a \quad \tau' = \pm \sigma_\tau$$

with an initial drift θ^0 and a final drift after the step θ^1. Since synthetic signals are used, filter performance can be evaluated through the comparison of $\hat{x}$ with actual x_r (which in real applications it is not usually known). The results can be seen in Fig. 5.4.

The basic cases correspond to:

1. Sensor perfectly known ($a' = a$, $\tau' = \tau$) with an step in the bias ($\theta^0 \neq \theta^1$). In this case, the higher K_∞ is, the faster the bias is corrected, as depicted in top plots of Fig. 5.4.
2. Constant bias and uncertainty on the sensor description ($a' \neq a$, $\tau' \neq \tau$). In this case, higher K_∞ values yield to rely excessively on the poorly known sensor dynamics, creating artificial peaks, as shown in bottom plots of Fig. 5.4.

The general error case in the sensor modelling is a (non-linear) superposition of the previous cases, and it is clear that a trade-off in the selection of K_∞ must be considered according to the signals uncertainties and the required convergence speed. For the sensors considered in this work, sometimes they exhibit a significant variation in their dynamic properties and a consistent model cannot be derived for the whole engine operating range. This would force a low value of K_∞ (and hence a slow correction of the model bias), but several modifications are proposed in the implementation.

Even though data based methods could be used for estimating the appropriate noise trade-off (Rajamani and Rawlings 2009), trial-and-error tuning is usual when working with KF. A Monte Carlo based method application presented in Payri et al. (2012) is proposed for tuning the filter, where the uncertainties on the sensor knowledge, expected working points (n and m_f values) and different expected θ are used for evaluating the filter performance under different σ_w^2/σ_v^2 values. These variations are considered with statistical distributions, and the configuration which minimises the total error is selected.

5.3 Learning Algorithms for Updating Look-up Tables

Look-up tables are often used in automotive in order to model systems with complex (and highly non-linear) expressions in a grid basis. The table elements are used for interpolating in order to generate the outputs. They can be included in a model, such

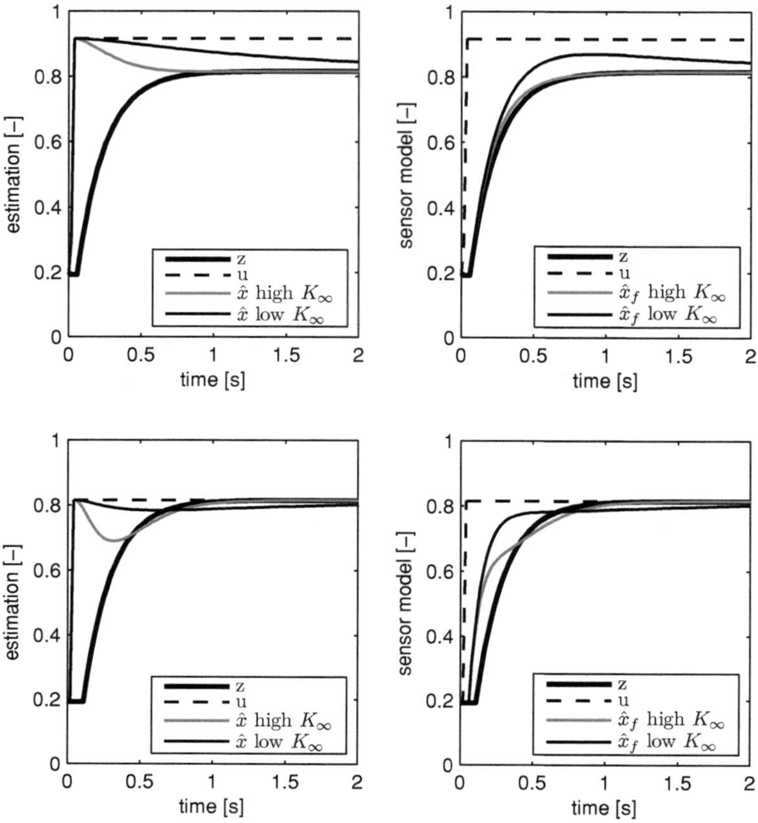

Fig. 5.4 *Top plots* Observation of the augmented model when the sensor model is perfectly known ($a' = a$, $\tau' = \tau$) with an step in θ. *Bottom plots* Observation of the augmented model when the sensor model is not perfectly known ($a' \neq a$, $\tau' \neq \tau$) and with a constant θ. The *left* plots show the model estimation $\hat{x}$ when the sensor performance is perfectly known and for two different σ_w^2/σ_v^2 quotients, i.e. two different values for K_∞. The *right* plots show the sensor model estimation $\hat{x}_f$ for these two K_∞ values

as the NO_x model presented in Chap. 4, or can be used directly for storing the bias θ calculated in the previous section. The possibility of developing RT algorithms for updating tables makes possible to develop adaptive look-up tables. The methods based on the EKF are well-suited but require a heavy computation burden, topic that is especially treated in this section. In the following, the section describes the problem of updating look-up tables and presents three methods based on the EKF (Guardiola et al. 2011, 2013a, b).

5.3.1 Defining Look-up Tables

A look-up table $\boldsymbol{\Theta} \in \mathbb{R}^{\prod_{i=1}^{N} n_i}$ is defined as a N-Dimensional mapping $\{\boldsymbol{\Theta} : \mathbb{R}^N \rightarrow \mathbb{R}\}$ represented by a grid in $N \in \mathbb{Z}^+$ dimensions, where each one has n_i grid points. The mapping further relies on a multivariate interpolation $q(\cdot)$ to calculate the function value from the input using the grid for the interpolation variables and $\boldsymbol{\Theta}$. In automotive systems, the multivariate interpolation schemes are often linear in each one of the dimensions, and this is the case that will be considered here. Without loss of generality the presentation will use 2D tables as being the most frequently occurring dimensions, since working with other dimensions is only a matter of reducing or increasing indexes. Then, for N = 2,

$$\boldsymbol{\Theta} = [\Theta_{i,j}] \tag{5.13}$$

where $i = 1, \ldots, n_r$ and $j = 1, \ldots, n_c$ (r stands for *row* and c for *column*). The multivariate interpolation function for generating the output $x_\Theta(k)$ from the input $u_k = [u_1(k)\ u_2(k)]^T$ can be expressed as

$$x_\Theta(k) = \text{vec}\,(q(u_k))^T\ \text{vec}(\boldsymbol{\Theta}) \tag{5.14}$$

vec(·) is the vector transformation and $q(u_k)$ is the interpolation matrix that both selects the elements to be interpolated and contains the weights. In the 2D case $q_k(u_k)$ selects the 4 (2^N in the ND case) active elements $\Theta_{i,j}$, $\Theta_{i,j+1}$, $\Theta_{i+1,j}$, $\Theta_{i+1,j+1}$ (with i, j fulfilling $u_{1,k} \in [r_i, r_{i+1}]$ and $u_{2,k} \in [c_j, c_{j+1}]$) and thus contains the following block with non-zero weights

$$\begin{bmatrix} q(u_k)_{i,j} & q(u_k)_{i,j+1} \\ q(u_k)_{i+1,j} & q(u_k)_{i+1,j+1} \end{bmatrix} = \begin{bmatrix} (1-\eta_{1,k})(1-\eta_{2,k}) & (1-\eta_{1,k})\eta_{2,k} \\ \eta_{1,k}(1-\eta_{2,k}) & \eta_{1,k}\eta_{2,k} \end{bmatrix} \tag{5.15}$$

where

$$\eta_{1,k} = \frac{u_{1,k} - r_i}{r_{i+1} - r_i}, \qquad \eta_{2,k} = \frac{u_{2,k} - c_j}{c_{j+1} - c_j} \tag{5.16}$$

This section analyses the computational aspects of updating look-up tables with the KF. Two algorithms are designed based on the KF for a computationally efficient table updating: a simplified KF (SKF) which manipulates the covariance matrix **P** and the associated updates efficiently, and a steady-state approach for the KF (SSKF) which directly neglects covariance information. Both methods are inspired by the works presented in Höckerdal et al. (2011), Vogt et al. (2004), Wu (2006).

5.3.2 Modelling for Learning, Drift or Ageing in Tables

In the setting, $\boldsymbol{\Theta}$ models a nonlinear function and the interesting aspect is to allow the model to adapt to the system to either learn the system and/or follow the ageing of the system. This is modelled in the standard way as a random walk process, where the table parameters are collected in a state vector $x_k \in \mathbb{R}^{n_r \times n_c}$. The full model can be written as follows

$$x_k = x_{k-1} + w_k \tag{5.17a}$$

$$z_k = \text{vec}(q(u_k))^T\, x_k + v_k \tag{5.17b}$$

noting that no uncertainties are allowed in the interpolation variable u_k. For convenience, the non-zero elements in $\text{vec}(q(u_k))$ are denoted $q_k^o \in \mathbb{R}^{1\times 4}$ (o stands for *observable*) and the corresponding elements in the state vector, $x_k^o \in \mathbb{R}^4$ for the 2D case. Then, the following expression is obtained for the output

$$z_k = q_k^o\, x_k^o + v_k \tag{5.18a}$$

$$q_k^o = \begin{bmatrix}(1-\eta_{1,k})(1-\eta_{2,k}) & \eta_{1,k}(1-\eta_{2,k}) & (1-\eta_{1,k})\eta_{2,k} & \eta_{1,k}\eta_{2,k}\end{bmatrix} \tag{5.18b}$$

where $\mathbf{Q}$ and $\mathbf{R}$ are

$$\mathbf{Q} = \sigma_w^2 \mathbf{I}_{\mathbf{n_r}\times\mathbf{n_c}} \tag{5.19a}$$

$$\mathbf{R} = \sigma_v^2 \tag{5.19b}$$

and $\mathbf{I}_{\mathbf{n_r}\times\mathbf{n_c}} \in \mathbb{R}^{\mathbf{n_r n_c}}$ is the identity square matrix. If there is application knowledge available, σ_w^2 might be selected individually for each table element building a vector $\Sigma_w^2 = [\sigma_{w,1}^2 \ldots \sigma_{w,n_r\times n_c}^2]$ that contains individual variances in such way $\mathbf{Q} = \Sigma_w^2 \mathbf{I}_{\mathbf{n_r}\times\mathbf{n_c}}$. In automotive, this is useful for considering the order of magnitude of the parameters, i.e. the absolute error will not be the same for low emissions at lower loads than highest peaks at higher loads, and on the other hand, if a driving pattern exists, the noise could be mapped over the table grid. This is linked with the foreseen probability that the engine is running at a certain operating condition, i.e. the ageing of the elements corresponding to more frequent areas will be lower and the opposite.

5.3.3 The Extended Kalman Filter, KF

The EKF (along the text and figures, the acronym KF refers to both EKF and KF) can now be used to observe x_k, when measurements z_k are given. At every iteration, the table is updated by

$$\hat{x}_k = \hat{x}_{k-1} + K_k(z_k - vec(q(u_k))^T \hat{x}_k) \tag{5.20}$$

where K_k is calculated using Eq. (2.13), $H_k = vec(q(u_k))^T$ varies with the operating point conditions. Although only the active elements x_k^o are updated at every k, all

P elements enter in the equation, which leads to huge calculations and big required memory resources. This makes difficult the implementation in commercial ECUs.

There are several publications on computational aspects of Kalman filters. Some authors have studied the filter optimization when different nonlinear functions are handled as for instance (Charalampidis and Papavassilopoulos 2011; Jørgensen et al. 2007; Singer and Sea 1971); where the latter makes an interesting study on the total number of operations required for the updating phase. Chandrasekar et al. (2007) present an interesting methodology when the system order is extremely large by using the finite-horizon optimization technique for obtaining reduced-order systems. But the key observation here is that the look-up table estimation has structural properties that can be exploited to reduce the computational and memory requirements significantly in a simple way, as suggested by Vogt et al. (2004).

To understand the problem, the covariance matrix **P** in the KF is studied. For instance, a 2D look-up table with a size $n_r n_c$ needs system **F** and covariance matrices **P** of $(n_r n_c) \times (n_r n_c)$. However, only the active elements are influenced during the update and by defining a local observable system (which corresponds to the active elements), one receives a system which is 4×4 against $(n_r n_c) \times (n_r n_c)$. That means that the non-active elements do not affect the updating until they become active (zero values for the K related elements). Furthermore, **P** is a symmetrical and positive-semidefinite matrix, which allows further simplifications in the KF calculations.

5.3.3.1 The Local Observable System

The *Local Observable System* is defined and analysed and a computationally efficient approximation for the KF is described (Guardiola et al. 2013b). Following with the 2D look-up table application, at every iteration a maximum number of 4 elements can be updated and then, if no dynamics are considered, the general $n_r n_c$ system (5.17a), (5.17b) can be simplified to

$$x_k^o = \mathbf{I}_{4\times4} x_{k-1}^o + w_k \tag{5.21a}$$

$$z_k = q_k^o x_k^o + v_k \tag{5.21b}$$

where $\mathbf{I}_4 \in \mathbb{R}^{4\times4}$ is the unitary matrix and $x_k^o \in \mathbb{R}^4$ is the state-vector that stores the active states at every k. The simplest 2D map is a table of only 4 elements, and where the one existing area is always active, i.e. the system (5.21a), (5.21b) is exactly (5.17a), (5.17b). For the general case, (5.21a), (5.21b) must be rewritten at every k as elements and matrices change.

5.3.3.2 About the System Observability

The system (5.17a), (5.17b) is not fully observable in one iteration. But the local observability is ensured if the parameters or states in the local system (5.21a), (5.21b) converge with a given data set. As the system is not LTI, the ordinary observability rank condition (Ogata 2001) is not directly applicable. q_k^o depends on the input data, and it is evident that if a enough level of excitation is given, then the system could be observed, whatever the method chosen. This minimum level of excitation could be proved if four elements have been excited. A sufficient observability matrix for local observability may be built with the first 4 independent observations not necessarily consecutive in instants i_1, i_2, i_3 and i_4 of the elements of the involved area:

$$\mathbf{O} = \begin{bmatrix} q_{i_1}^o & q_{i_2}^o & q_{i_3}^o & q_{i_4}^o \end{bmatrix}^T \tag{5.22}$$

and if the rank of this matrix is 4, then the system (5.21a), (5.21b) might be observable. But here the problem is linked with the noise tuning and indeed full rank $\mathbf{O}$ does not lead to system full convergence (full observability does not lead to full convergence as the method and the model structure affect). The most of the elements are also included in other neighbour areas and because of this, the minimum condition of observability of a given state depends on the number of independent measurements that affects every state. The condition stated in Höckerdal et al. (2011) gives a general condition.

5.3.4 The Steady-State KF Approach, SSKF

If the problem is considered stationary (the same input is repeated once and again), then $q_k^o = q^o$ and the filter is steady-state (Guardiola et al. 2013a). In addition to this and given a noise ratio σ_v^2/σ_w^2, K_k in (2.13) converges after a certain number of iterations as expressed in (5.9). That occurs although $\mathbf{P}$ does not converge and can be ill conditioned. Then, the stationary equivalent filter is obtained and an analytical expression is derived for the *local observable system*, resulting a Kalman gain $K^{ss} \in \mathbb{R}^4$

$$K^{SS} = \begin{bmatrix} k_i(\eta_1, \sigma_v^2/\sigma_w^2) \cdot k_i(\eta_2, \sigma_v^2/\sigma_w^2) \\ k_i(\eta_1, \sigma_v^2/\sigma_w^2) \cdot k_i(1-\eta_2, \sigma_v^2/\sigma_w^2) \\ k_i(1-\eta_1, \sigma_v^2/\sigma_w^2) \cdot k_i(\eta_2, \sigma_v^2/\sigma_w^2) \\ k_i(1-\eta_1, \sigma_v^2/\sigma_w^2) \cdot k_i(1-\eta_2, \sigma_v^2/\sigma_w^2) \end{bmatrix} \tag{5.23}$$

where k_i is computed as follows

$$k_i(\eta, \sigma_v^2/\sigma_w^2) = \frac{0.5(1-\eta)(1+s)}{0.5(1+s)(1-2\eta+2\eta^2) + \sigma_v^2/\sigma_w^2} \tag{5.24a}$$

$$s = \sqrt{1 + \frac{4}{(1-2\eta+2\eta^2)} \frac{\sigma_v^2}{\sigma_w^2}} \tag{5.24b}$$

These expressions are solved in Appendix A.3.2.

The user can evaluate (5.23) for the 2D case for a given σ_v^2/σ_w^2 or can simulate the functions off-line for mapping the values and reduce the computing time, which is the

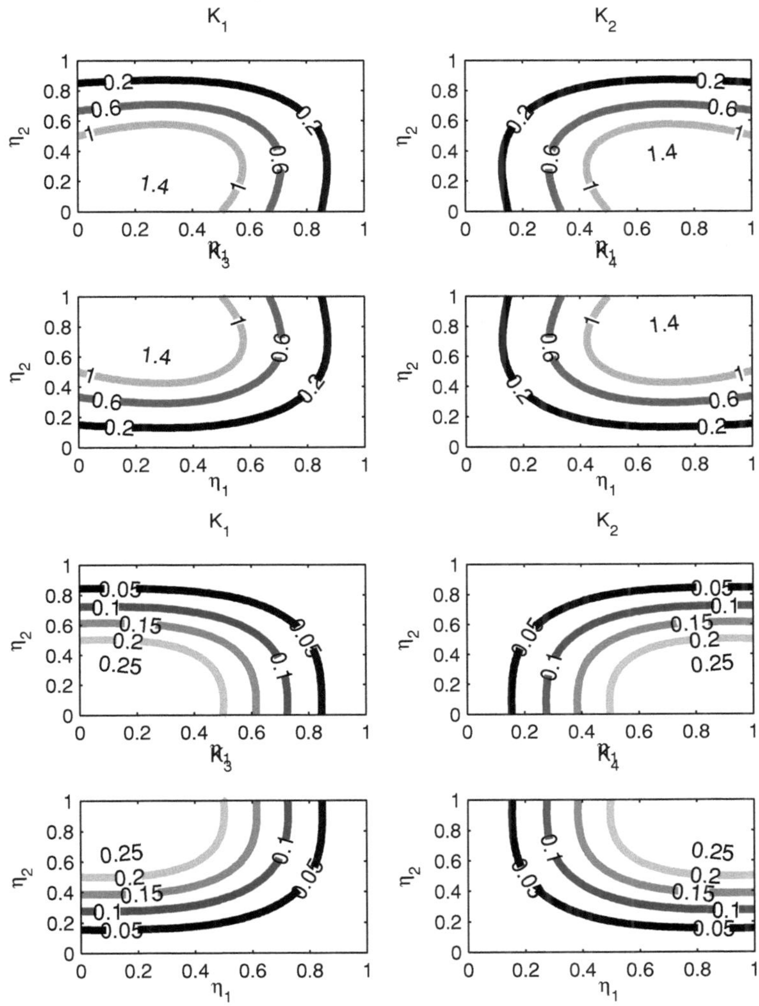

Fig. 5.5 Membership functions for Kalman gain correction for the 2D case for $\sigma_v^2/\sigma_w^2 = 0$ (*top*) and $\sigma_v^2/\sigma_w^2 = 0.1$ (*bottom*)

option suggested by the authors. Figure 5.5 shows the values of the K^{SS} elements for two different σ_v^2/σ_w^2 quotients. The symmetry of the functions is clear when comparing the elements of K^{SS}. Both the noise tuning and the relative position of the operating points affect to the estimation. On one hand, as the operating point approaches to one grid node, the gain is closer to the maximum (it is not linear) and the opposite. On the other hand, a higher σ_v^2/σ_w^2 leads to a higher correction. Non-active elements have a null correction, i.e. their related gain is zero.

This approach allows to map K^{SS} and this only depends on the inputs [η_1 and η_2 from (5.16)] and the noise trade-off σ_v^2/σ_w^2. Nevertheless, as the parameters ageing is not considered nor the estimation error computed, the algorithm is quite fast and light but the robustness must be assessed in the applications. The pseudo-code of the algorithms is in the Appendix A.4. The next subsection presents the SKF algorithm which tracks the estimation error in an efficient way.

5.3.5 The Simplified Kalman Filter, SKF

Despite that only the associated active elements affect K_k, all $\mathbf{P_k}$ elements are predicted and updated at every k. $\mathbf{P_k}$ can be reorganised in such way that variances related with observable $\mathbf{P_k^o}$ and unobservable $\mathbf{P_k^u}$ elements are split.

$$\mathbf{P_k} = \left[\begin{array}{c|c} \mathbf{P_k^u} & \vdots \\ \hline \cdots & \mathbf{P_k^o} \end{array}\right] \tag{5.25}$$

K_k, q_k and $\mathbf{Q}$ can also be split in the same way. Iterating one step ahead (2.13c) is written

$$\begin{aligned} \mathbf{P_{k+1}} &= \left[\begin{array}{c|c} \mathbf{I} & \mathbf{0} \\ \hline \mathbf{0} & \mathbf{I_{4\times4}} - K_k^o q_k^o \end{array}\right] \left[\begin{array}{c|c} \mathbf{P_k^u} + \mathbf{Q^u} & \vdots \\ \hline \cdots & \mathbf{P_k^o} + \mathbf{Q^o} \end{array}\right] \\ &= \left[\begin{array}{c|c} \mathbf{P_k^u} + \mathbf{Q^u} & \vdots \\ \hline \cdots & (\mathbf{I_4} - K_k^o q_k^o)(\mathbf{P_k^o} + \mathbf{Q^o}) \end{array}\right] \end{aligned} \tag{5.26}$$

where all observable sub-matrices are 4×4, while the unobservable ones are $(n_r n_c - 4) \times (n_r n_c - 4)$. (5.26) shows how the diagonal elements of $\mathbf{P_k^u}$ are only affected by adding the diagonal matrix $\mathbf{Q^u}$. However, both diagonal and non-diagonal elements of $\mathbf{P_k^o}$ are affected as in a 4×4 KF. The crossed relationships between the observable and the unobservable system, shown as dots in the matrix, are kept constant. This allows simplifying the complete system to the 4×4 active state-space system, whose resolution is highly computationally efficient.

The current work develops an approximation of the KF that requires both less memory and computations in the iterations and is named SKF. The SKF builds and solves the local observable system (5.21a), (5.21b) at every k. Some simplifications are assumed: for the non-active elements, whose covariance matrix is $\mathbf{P^u}$, non-diagonal elements are neglected building a vector $P^u = diag(\mathbf{P^u})$. This is justified for two reasons. First, when the system is running, then leaves one area and later returns to it, related variances reflect the time that the system has been out of this area. This is a desired property since the ageing is captured by the variance increase due to $\mathbf{Q^u}$, which indeed for this application it is defined as a diagonal matrix. Older crossed correlations are maintained over the time and whether or not they are valid is not sure. Second, the simulation results show that the full EKF and the SKF have similar performances when running on synthetic data and real data from the engine application. Then, at every iteration, the local system (5.21a), (5.21b) is solved with the only additional operation of $P_k^u + Q^u$ (converted to a vector calculation). The SKF is a suboptimal filter but with a similar behaviour than KF, as demonstrated in Sects. 5.3.6.

The pseudo-code for the SKF implementation is in Algorithm A.4. Note that the code is to show the algorithm where some variables are not strictly needed in the computer programming, e.g. P_k^u always that P_k exists, but these steps are kept for the sake of clarity. The noise variance σ_w^2 is selected equal and constant for all elements, allowing to pre-define a diagonal matrix $\mathbf{Q^o} = \sigma_w^2 \mathbf{I_{4\times 4}}$ for the active system and a vector $Q^u = \sigma_w^2 diag\left(\mathbf{I_{n_r \times n_c - 4}}\right)$ for the non-active one.

In the next, the memory allocation and computational complexity of the three methods, full EKF (here denoted KF), the SKF as well as the SSKF are described.

5.3.5.1 Analysis of the Memory and Computations

A computation time study is made programming code in the Matlab software on a laptop computer equipping Intel Core 2DUO T9300 2.5 GHz with Windows VISTA 64 bits. Figure 5.6 shows the relative computation time used by the three methods for updating 2D look-up tables of different sizes. Only square $n \times n$ tables have been considered with n ranging from 2 to 21, since it is the total number of elements in $vec(T)$ that influences memory allocation and computations. The Y-axis shows the relative computation time with respect to the averaged time that the SSKF takes for performing 1,000 iterations, i.e., a relative time of 1 means that the method needs exactly the same time than the SSKF. The SKF needs around 1.15 times the SSKF calculation time, whatever the number of parameters. This slight difference is explained by the manipulation of variances and the need of redefining the active vector and matrices $\mathbf{P^o}$, and x^o when the active area changes (see lines 4 and 5 of the Algorithm 2 and compare with Algorithm 1, both included in Appendix A.4). However, the required time for the KF rapidly grows when the number of parameters increases, i.e. for 256 parameters (16 × 16), the KF needs around 7.8 times the one for the SSKF. An exponential function may be fitted by the least squares method to the KF time as function of the number of parameters n_p, whose result is the relative time that the KF takes with respect to the one of the SSKF.

Table 5.1 summarises the required memory resources and computation times for the algorithms. The SKF permits to reduce the requirements in terms of memory resources and the general system $(n_r n_c) \times (n_r n_c)$ is reduced to a simple 4×4. These results show the benefits when using the SSKF and the SKF compared to the full KF.

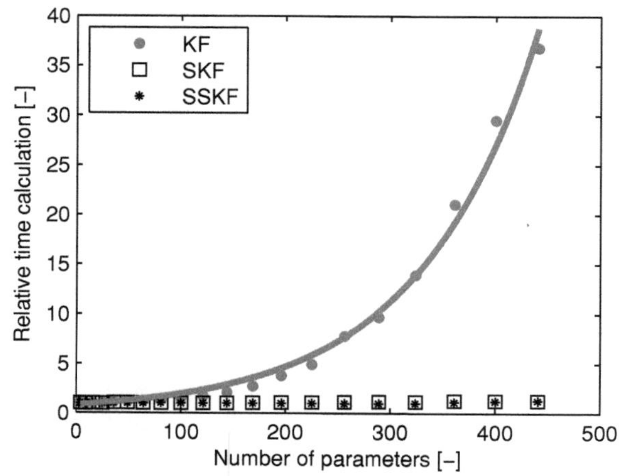

Fig. 5.6 Relative computation time required for the methods for updating 2D look-up tables. Times are normalised with the averaged time that the SSKF needs for computing 1000 iterations of the filter

Table 5.1 Memory storage and computational burden comparison between the Kalman filter KF, the simplified Kalman filter SKF and the steady-state approach SSKF

	P	**K**	Rel. time
KF	$(n_r n_c) \times (n_r n_c)$	$(n_r \times n_c) \times 1$ (by 2.13)	$0.82\exp(0.0087 n_p)$
SKF	$4 \times 4 + (n_r \times n_c) \times 1$	4×1 (by 2.13)	1.15
SSKF	4×4	4×1 (by 5.23)	1

5.3.6 Simulation of the Updating Algorithms

The convergence and robustness of the approximations need also to be studied and this is first performed by simulations. The objective of this section is comparing the abilities and performance of the algorithms against different synthetic cases where no dynamics are accounted (the real system and the model are statics; Sect. 5.3.7 later discusses how to consider the dynamics).

For automotive applications, engine speed n and fuel mass injection quantity m_f usually define the engine operating point and the look-up table scheduling points selected in the simulation are inspired by these quantities. Hence r represents a grid for $u_1(k)$ (the speed in rpm) and c for $u_2(k)$ (the injected fuel mass in mg/str)

$$r = [500 \quad 1500 \quad 2500 \quad 3500] \tag{5.27a}$$

$$c = [0 \quad 20 \quad 40 \quad 60] \tag{5.27b}$$

and the true map (the objective map to update) $\mathbf{\Theta}_r$ is defined with the surface

$$\mathbf{\Theta}_r = \begin{bmatrix} 1 & 1 & 1 & 1 \\ 1 & 1 & 1 & 1 \\ 1 & 1 & 1 & 1 \\ 1 & 1 & 1 & 1 \end{bmatrix} \tag{5.28}$$

where $x_r = vec(\mathbf{\Theta}_r)$ is the expanded state vector objective of the estimation process. No initial process knowledge is considered: x_0, x_0^o, $\mathbf{P_0}$ are initialised with zeros. $\sigma_v^2/\sigma_w^2 = 1$ for all cases, unless it were pointed.

System measurements are given and the algorithms performance is evaluated. As synthetic signals are used, this allows defining error metrics. Simulations include two worst-cases and a favourable one:

- Input with random variation: a random shot of values following an uniform probability distribution. This is an ideal situation for learning as all areas are excited and the parameters ageing is low (the excitation is high). Measurements are perfect and constant, thus $z_k = 1\ \forall k$.
- Linear variation: varying u_2 keeping constant u_1. This variation tests a degenerated case where observability is critical along the u_1-dimension. Measurements are perfect and constant: $z_k = 1\ \forall k$.
- Measurement noise rejection: studying the effects of noise transmission between output and observation when the measurement vector of the previous case is noisy.

5.3.6.1 Simulation 1: Input with Random Variation

The sequence $\{u_k\}_{k=1}^{1000}$ following an uniform distribution over the grid defined in (5.27a), (5.27b) is used for exciting the system. A complete identification is possible as the grid is completely covered. This situation is not so far of the reality for diesel engines, e.g. urban cycles with a lot of speeding/braking actions cause quasi-random variations in partial-low load areas of the engine map. The selected grid and covered points are shown in Fig. 5.7.

In this case, the variance information is not as relevant as in other cases, because of the stochastic nature of inputs. The mean value of all states is plotted at each time step in Fig. 5.8. All three methods perform well and the rate of convergence can be tuned varying σ_v^2 and σ_w^2.

A quick view on the effects of the hypothesis for the SKF is shown for one element x_6 in top plot of Fig. 5.9, which shows the variance P_6 of x_6 for the KF and the SKF (not applicable for the SSKF). This value indicates the observability of x_6: when it increases monotonically, x_6 is not observable. In some parts, an offset between P_6 for the KF and the SKF appears because of the neglected covariances in the SKF, but in the end, this offset is absorbed and the two methods behave similar. The bottom plot of Fig. 5.9 shows x_6 evolution for the three methods. x_6 is updated when the element becomes observable. Note that x_6 is already active in the first iteration and K_1^{SS}, which represents the converged value of the equivalent LTI system, is non-null. Then, the SSKF updates x_6 in $k = 1$ while K_1 for the KF and the SKF is null.

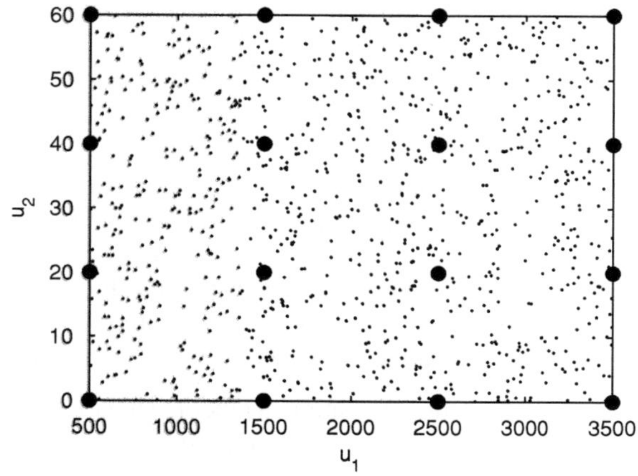

Fig. 5.7 Random variation of inputs u_1 and u_2 shown with *small dots* and table grid with *thick dots*

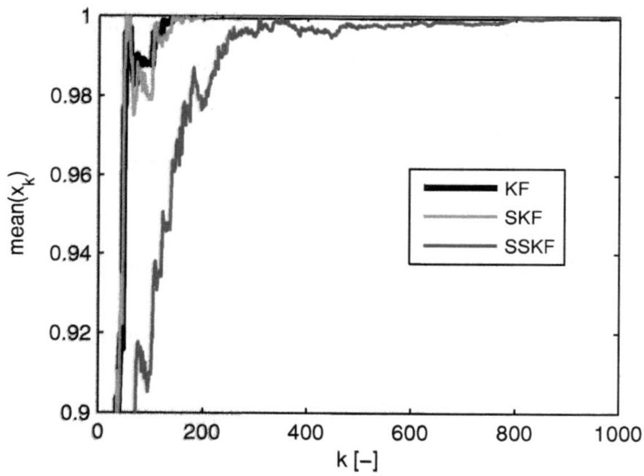

Fig. 5.8 Averaged value of table parameters for the three updating methods when considering an uniform distributed inputs sequence. The correct value is 1

Anyway, this is not an advantage of this method: the KF and the SKF can behave similar if P_0 is non-null, but here for the simulations $\mathbf{P_0} = [\mathbf{0}]$. When there exists no previous knowledge of the system to be learnt, it should advisable to initialize P_0 with a certain value for speeding up the updating during the first iterations.

5.3.6.2 Simulation 2: Input with a Linear Variation

$\boldsymbol{\Theta}_r$ is identified with a sequence $\{u_k\}_{k=1}^{200}$ where u_1 varies monotonically from 0 to 60 (from $k = 0$ to 100) and coming back from 60 to 0 (from $k = 101$ to 200), while $u_1 = 2{,}000$ $\forall k$. This condition is highly restrictive as the excitation level is not high: at each k only small variations of input u_2 are applied and due to this, the full convergence condition is not fulfilled in the first running of one area, and the

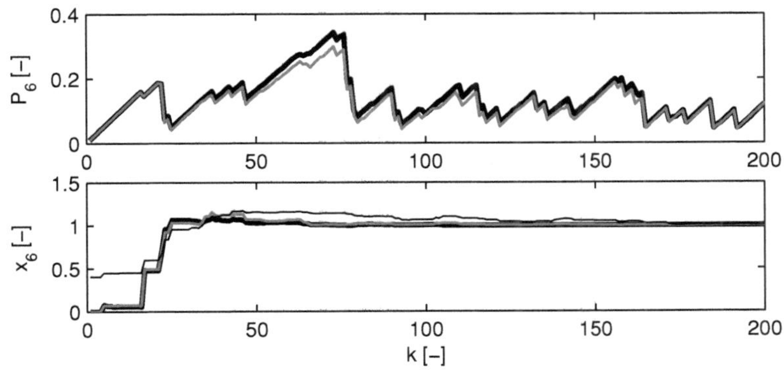

Fig. 5.9 Variance analysis for an uniform distributed inputs sequence. *Top* Variance evolution of the 6th element P_6 where *black thick line* is the KF and *gray line* is the SKF (P_6 is normalised by 100 in both cases). *Bottom* x_6 value where *black thick line* is the KF, *gray line* is the SKF, *thin black* is the SSKF

interactions between areas and the way back are necessary to ensure the convergence of both the KF and the SKF. Three matrix areas are excited [see (5.29)]; area 1: elements 5, 6, 9, 10; area 2: 6, 7, 10, 11; area 3: 7, 8, 11, 12. Moreover, due to the inputs nature, the row 5–8 evolution is equivalent to that of 9–12.

$$\boldsymbol{\Theta}_k = \left[\begin{array}{c|c|c|c} \cdot\,\cdot & & \cdot & \cdot \\ \hline x_{5,k} & x_{6,k} & x_{7,k} & x_{8,k} \\ x_{9,k} & x_{10,k} & x_{11,k} & x_{12,k} \\ \hline \cdot\,\cdot & & \cdot & \cdot \end{array} \right] \tag{5.29}$$

Figure 5.10 shows the evolution of the 4 states x_5 to x_8. First thing to note is the similar performance of the KF and the SKF and the instabilities of the SSKF. For instance, x_5 is in the area 1, and during the first run it is not capable of converging, but in one complete round the KF and SKF methods converge, and the SSKF seems to do it, at least in $k = 200$. The other states have slightly different performances, because the first area has been already been covered; e.g. x_6 is closer to the convergence as being a member of area 1 and area 2, but until the way round it does not get the full convergence for the KF and SKF methods. Similar behaviours are observed for x_7 with these methods. In addition, x_8, as member of the area 3, has the advantage that areas 1 and 2 have been covered first, and the information is already available for getting the full convergence for the KF and SKF in $k = 100$. The full observability (in the sense of convergence) of these 8 elements is reached only when $\mathbf{O}$ in (5.22) copes with the three areas when using the KF and SKF. Nevertheless, the SSKF estimation does not converge for x_6, x_7 and x_8 due to no state error information is tracked, and the system and convergence becomes more dependent on data.

In order to understand how KF and SKF behave, it is informative to also study the variance and covariance evolution (remember that this is not present for the

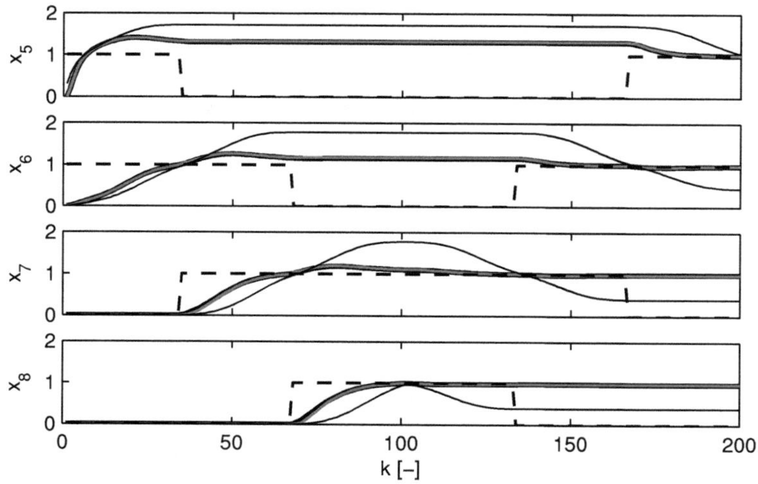

Fig. 5.10 States evolution for a monotonically varied inputs sequence. *Black thick line* is KF, *gray line* is SKF, *thin black* is SSKF and *dashed line* shows when the element is observable (1 is active; 0 is not active)

SSKF). Figure 5.11 shows the variance evolution for row elements, while Fig. 5.12 shows the evolution of a few selected covariances between states (corresponding to non-diagonal elements of **P**). As for the random case in Fig. 5.9, variance evolution for the KF and SKF is similar, although some deviation exists in elements 6 and 7. This is explained for covariances or non diagonal $P_{ij}\ \forall i \neq j$ elements that the SKF only accounted for active areas. This is shown in Fig. 5.12 where some crossed covariances are plotted. When the pair of elements is not active, covariances of these are not tracked, equivalent to reset them to zero in the global **P**. This is the particular difference with the KF, which always tracks the covariance, although the area would not be active, increasing the computing requirements without a justified improvement in the estimation. However, when an area is active, the SKF covariance evolves as the KF and in the limit is equivalent to the KF. The difference is larger initially but after the observations have converged the difference is negligible, as shown in Fig. 5.10. This pattern may slightly be modified linking the covariance values to elements pairs and not to the involved areas (excepting boundaries, all pair of elements are shared in two different areas), although final results are similar.

5.3.6.3 Simulation 3: Measurement Noise Rejection

An uniform distributed noise with zero mean and maximum amplitude of 0.2 is applied to the measurement given in simulation 2; z_k is shown in Fig. 5.13. A robust method must filter this noise and converges to the true values.

The resulting estimations are shown in Fig. 5.14, and the evolutions are quite similar to the ones of Fig. 5.10 although with a slight noise transmission. The KF

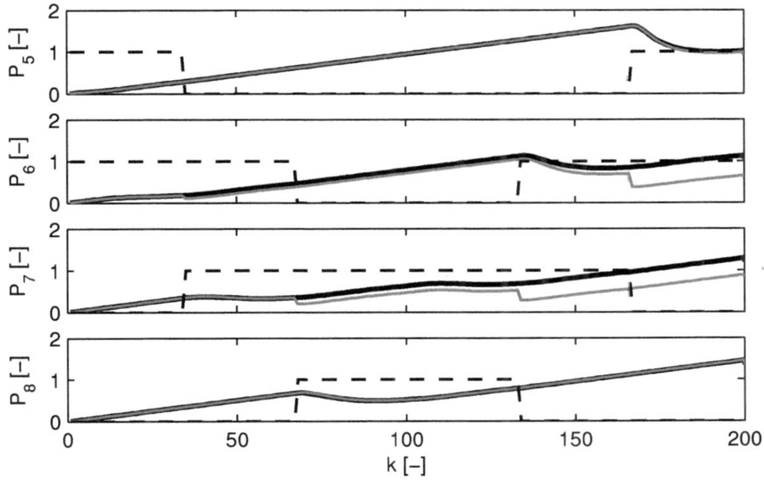

Fig. 5.11 P_{ii} analysis for a monotonically varied inputs sequence. Values are normalised by 100. Variances of elements 9–12 are equal to those of 5–9 because of symmetric properties. *Black line* is KF, *gray line* is SKF and *dashed line* shows when the element is active (1 active; 0 is not active)

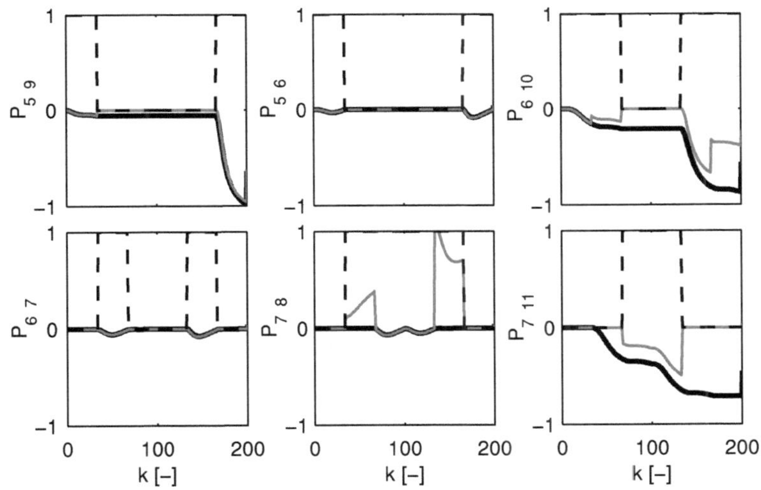

Fig. 5.12 P_{ij} analysis for a monotonically varied inputs sequence. Values are normalised by 100. *Black line* is KF, *gray line* is SKF and *dashed line* shows when the element is active (1 active; 0 is not active)

and SKF perform well, being able to also filter the noise, but the SSKF is not capable of converging again having a similar response as in the simulation 2. Variance and covariances are exactly the same of Figs. 5.11 and 5.12 as the state-space system and the inputs sequence does not change.

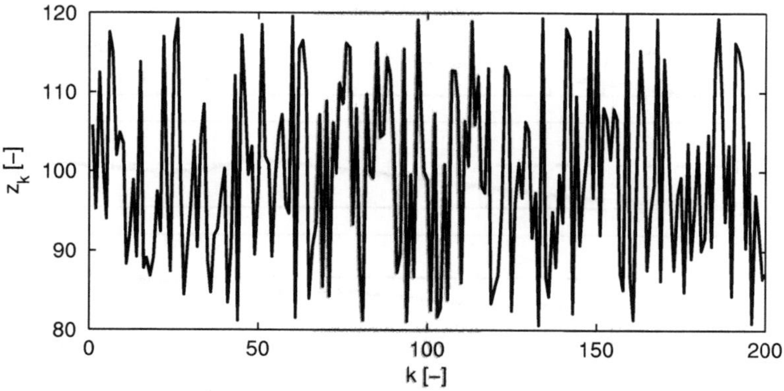

Fig. 5.13 Measurement z_k plot with the addition of a uniform distributed noise with zero mean and maximum amplitude of 0.2

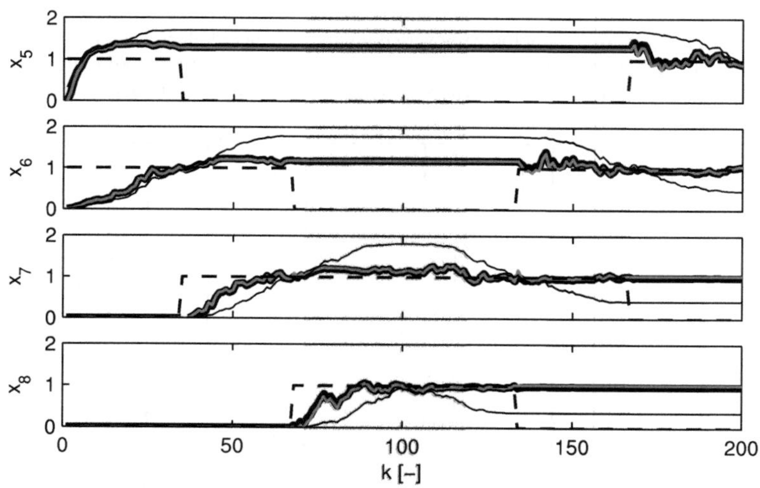

Fig. 5.14 States evolution for noisy measurements. *Black thick line* is KF, *gray line* is SKF, *thin black* is SSKF and *dashed line* shows when the element is active (1 is active; 0 is not active)

5.3.6.4 Conclusions from the Simulations

The SKF method is demonstrated to have similar accuracy as the KF, at least for the numerical cases proved in this section. Furthermore, the KF and SKF solutions converge for all cases, including the simulations 2 and 3, which are specifically restrictive because of the low level of excitation. The KF, as shown in Fig. 5.6 and Table 5.1, is computationally heavy. The SSKF is the fastest and lightest method as the variance is not tracked and K^{SS} formulation is derived analytically. The SKF only needs around 1.15 times the SSKF time for calculation, and the required memory resources are similar, except for the track of the variances. The SSKF behaves well in situations where the variance tracking is not critical, e.g. the simulation 1, but it is less robust when the data is structured and the level of excitation is low, e.g. simulations 2 and 3. The KF and SKF are also capable of filtering the noise, a minimum condition for a correct updating, and SSKF also filters the noise, but its stability depends on the data. In comparison, the SKF method, derived from the KF, is demonstrated to be an efficient and accurate method for updating look-up tables, at least for the presented simulation conditions.

5.3.7 The Dynamic Equations for Learning

The learning algorithms presented in the previous subsections have considered a quasi-static model, i.e. the model outputs are directly calculated by a table interpolation. In this subsection, the state-space model (5.17a), (5.17b) is slightly modified for coping with sensor dynamics and inputs u are treated as well.

Usual inputs for the tables in diesel engines are the signals of the ECU: speed n and injection m_f, and as far as they are fast signals, no special treatment is necessary in principle, while the delays are still needed to phase the system correctly with the NO_x sensor

$$u_{1,k} = z^{-\tau/T_s}\, n(k) \tag{5.30a}$$

$$u_{2,k} = z^{-\tau/T_s}\, m_f(k) \tag{5.30b}$$

where τ represents the averaged sensor delay obtained with the procedures shown in Chap. 3 (T_s is the sampling period). This is preferred to apply the delay in the state-space model for avoiding to increase the dimensions of the system.

In this basis, the table output $x_\Theta(k) \in \mathbb{R}$ is calculated by means of interpolating the adaptive map $\mathbf{\Theta}_k$

$$x_\Theta(k - \tau/T_s) = \mathbf{\Theta}_k(u_{1,k}, u_{2,k}) \tag{5.31}$$

defining a quasi-static representation of the table output where k defines the current time.

Sensor dynamics are considered as in (3.3) for computing $x_{\Theta f}(k) \in \mathbb{R}$

$$x_{\Theta f}(k) = \frac{1-a}{1-az^{-1}} x_\Theta(k - \tau/T_s) \tag{5.32}$$

and the global system (5.17a), (5.17b) must be augmented with one extra-dimension considering $x_{\Theta f}(k)$

$$x_k^w = \left[\begin{array}{c|c} \mathbf{I_{n_r n_c \times n_r n_c}} & \\ \hline (1-a)q_k & a \end{array}\right] x_{k-1}^w + w_k \tag{5.33a}$$

$$z_k^w = H^w x_k^w + v_k \tag{5.33b}$$

with

$$x_k^w = \begin{bmatrix} x & x_{\Theta f} \end{bmatrix}_k^T \tag{5.34}$$

$$H^w = [0 \cdots 1] \tag{5.35}$$

being $x_k^w \in \mathbb{R}^{n_r n_c + 1}$ with $x_k \in \mathbb{R}^{n_r n_c}$ as in (5.17a), (5.17b), and $z_k^w \in \mathbb{R}$. The Index w stands for *wide*.

The local observable system (5.21a), (5.21b) is now

$$x_k^{ow} = F_k^{ow} x_{k-1}^{ow} + w_k \tag{5.36a}$$

$$z_k^{ow} = H^{ow} x_k^{ow} + v_k \tag{5.36b}$$

$$F_k^{ow} = \left[\begin{array}{c|c} & 0 \\ \mathbf{I_{4\times4}} & 0 \\ & 0 \\ & 0 \\ \hline (1-a)q_k^o & a \end{array}\right] \tag{5.37}$$

$$H^{ow} = [0\ 0\ 0\ 0\ 1] \tag{5.38}$$

$$x_k^{ow} = \begin{bmatrix} x_1^o & x_2^o & x_3^o & x_4^o & x_{\Theta f} \end{bmatrix}_k^T \tag{5.39}$$

where H^{ow} is constant, because q_k^o and the filtering parameter a are included now in the time varying process matrix $\mathbf{F_k^{ow}}$. $\mathbf{Q^{ow}}$ process noise matrix is also augmented, and includes an extra-noise term σ_f^2 for the new state $x_{\Theta f}$

$$\mathbf{Q^{ow}} = \left[\begin{array}{c|c} \mathbf{Q} & \mathbf{0} \\ \hline \mathbf{0} & \sigma_f^2 \end{array}\right] \tag{5.40}$$

Even though the new local system has one more dimension (5×5), the SKF hypotheses are still applicable as proved in Appendix A.5. Chapter 6 will prove the algorithms under real conditions in a diesel engine considering dynamics (the sensor output will be used there as reference) and completing the study on the methods capabilities.

5.4 Conclusions

This chapter has presented different observers based on the KF in order to estimate engine variables and update look-up tables.

First, a drift correction algorithm is proposed by augmenting a state-space model with an extra-state for tracking the drift which exists between a fast biased model and a delayed and filtered but steady-state accurate sensor signal. The data fusion is done through a steady-state KF, which may be computed beforehand. The algorithm allows to estimate online the actual considered signal.

Second, the attention turns out to the design of adaptive algorithms for look-up tables fitting and updating. Kalman filtering is used for updating because of its capability for tracking the system and parameters ageing. Computational issues involved when of updating look-up tables online with the KF based methods are then addressed. Two novel methods are designed for the current work:

- A steady-state KF (SSKF) approach using precalculated membership functions based on an off-line solution to the KF main equation. Table elements are updated in a fast way and the computational burden is low. The method relies on neglecting the covariance information, given by the standard KF, and then setting out the stationary problem where the inputs can be considered as constants in order to calculate the steady-state gain.
- A simplified KF (SKF) with similar accuracy as the standard KF, but that requires a much lower memory resources and calculation time. The local observable system serves as inspiration for analysing the variance matrix performance and is used for comparison. The SKF is based on the KF but neglects covariances of locally unobservable states.

The only requirement for the full table adaptation when using online learning algorithms is running the engine in all the matrix areas. The stability and convergence is assessed by means of simulating the algorithms. The memory saving and computational reduction that SKF offer, as well as the robust performance, make this algorithms suitable for being implemented on commercial ECUs. The SSKF has also an interesting performance and a very low computational burden, although the stability depends on the data quality.

In the following chapter, these algorithms are combined and different results are presented in order to estimate NO_x and λ^{-1} in diesel engines.

References

Alberer D, del Re L (2009) Fast oxygen based transient diesel engine operation. SAE technical paper 2009–01-0622

Chandrasekar J, Kim IS, Bernstein DS (2007) Reduced-order Kalman filtering for time-varying systems. Proceedings of the 46th IEEE conference on decision and control. LA, USA, New Orleans, pp 12–14

Charalampidis AC, Papavassilopoulos GP (2011) Computationally efficient Kalman filtering for a class of nonlinear systems. IEEE Trans Autom Control 56(3):483–491
Desantes JM, Luján JM, Guardiola C, Blanco-Rodriguez D (2011) Development of NO_x fast estimate using NO_x sensors. In: EAEC 2011 congress, Valencia
Galindo J, Serrano JR, Guardiola C, Blanco-Rodriguez D, Cuadrado IG (2011) An on-engine method for dynamic characterisation of NO_x concentration sensors. Exp Therm Fluid Sci 35(3):470–476
Guardiola C, Pla B, Blanco-Rodriguez D, Cabrera P (2011) Adaptive modelling for automotive applications. In: Proceedings of modelling for engineering and human behaviour 2011, Valencia. Universitat Politècnica de València
Guardiola C, Pla B, Blanco-Rodriguez D, Cabrera P (2013a) A learning algorithm concept for updating look-up tables for automotive applications. Math Comput Modell 57(7–8):1979–1989
Guardiola C, Pla B, Blanco-Rodriguez D, Eriksson L (2013b) A computationally efficient Kalman filter based estimator for updating look-up tables applied to NO_x estimation in diesel engines. Control Eng Pract 21(11):1455–1468
Guardiola C, Pla B, Blanco-Rodriguez D, Mazer A, Hayat O (2013c) A bias correction method for fast fuel-to-air ratio estimation in diesel engines. Proc Inst Mech Eng Part D: J Automobile Eng 227(8):1099–1111
Höckerdal E, Frisk E, Eriksson L (2011) EKF-based adaptation of look-up tables with an air mass-flow sensor application. Control Eng Pract 19:442–453
Jørgensen JB, Homsen PG, Madsen H, Kristensen MR (2007) A computationally efficient and robust implementation of the continuous-discrete extended Kalman filter. In: Proceedings of the American control conference
Kalman RE (1960) A new approach to linear filtering and prediction problems. J Basic Eng 82:35–45
Kalman RE, Bucy RS (1961) New results in linear filtering and prediction theory. J Basic Eng 83:95–108
Ogata K (2001) Modern control engineering, 4th edn. Prentice Hall, Boston
Payri F, Guardiola C, Blanco-Rodriguez D, Mazer A, Cornette A (2012) Methodology for design and calibration of a drift compensation method for fuel-to-air ratio estimation. SAE technical paper 2012–01-0717
Rajamani MR, Rawlings JB (2009) Estimation of the disturbance structure from data using semidefinite programming and optimal weighting. Automatica 45:142–148
Schilling A, Amstutz A, Guzzella L (2008) Model-based detection and isolation of faults due to ageing in the air and fuel paths of common-rail direct injection diesel engines equipped with a λ and a nitrogen oxides sensor. Proc Inst Mech Eng Part D: J Automobile Eng 222:101–117
Schilling A, Amstutz A, Onder CH, Guzzella L (2006) A real-time model for the prediction of the NO_x emissions in DI diesel engines. In: Proceedings of the 2006 IEEE international conference on control applications, Munich, Germany
Simon D (2001) Kalman filtering. Embed Syst Program 14(6):72–79
Singer R, Sea RG (1971) Increasing the computational efficiency of discrete Kalman filters. IEEE Trans Autom Control 16(3):254–257
Vogt M, Mnller N, Isermann R (2004) On-line adaptation of grid-based look-up tables using a fast linear regression technique. J Dyn Syst Meas Control Trans ASME 126(4):732–739
Wu G (2006) A table update method for adaptive knock control. SAE technical paper 2006–01-0607

Chapter 6
Adaptive Estimation of NO_x and λ^{-1}

6.1 Introduction

This chapter uses the adaptive algorithms developed in Chap. 5 in order to estimate λ^{-1} and NO_x, when a NO_x sensor is available. The use of adaptive algorithms also allows to tune model parameters with and without an initial knowledge (Schilling 2008), i.e. the algorithms can be used for online model fitting. For that purpose, three different algorithms are designed and the filter gains are calculated on the basis of the methods presented in Chap. 5:

- Algorithm A: The drift correction algorithm corrects the bias in the λ^{-1} and NO_x models. The bias is tracked online but models are not re-calibrated.
- Algorithm B: An adaptive table is used for learning and prediction without requiring the RT NO_x model. This approach is valid for slow cycles, such as the NEDC. The table is updated by using the SKF or the SSKF methods, as advisable.
- Algorithm C: Model is adapted and re-tuned online by tracking the nominal NO_x. With this strategy, the nominal NO_x map can be fitted by using the dynamic tests, indeed avoiding the steady-state campaign usually needed for its calibration. The state-space system (5.17) built for updating look-up tables should be slightly modified in order to consider the model structure.

Figure 6.1 schematises the three options, particularising for the NO_x estimation. The objective is on one hand, obtaining a fast and reliable estimation of the considered variable, i.e. without sensor filtering and delay and, on the other hand, introducing adaptive capabilities to the models.

In the following, these algorithms are selected for the fast estimation of λ^{-1} and NO_x.

D. Blanco-Rodriguez, *Modelling and Observation of Exhaust Gas Concentrations for Diesel Engine Control*, Springer Theses, DOI: 10.1007/978-3-319-06737-7_6,

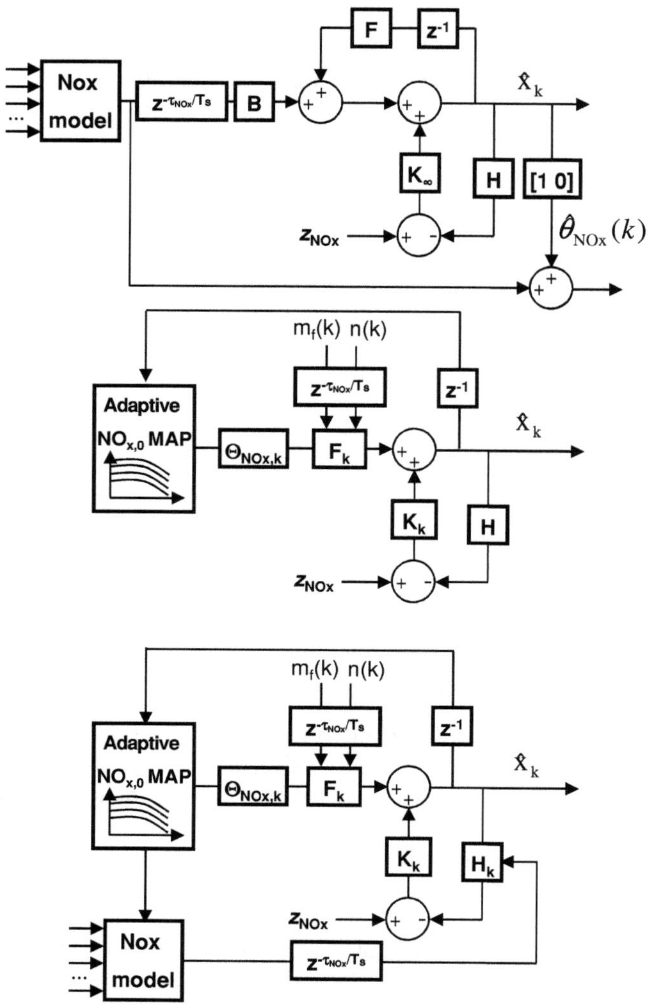

Fig. 6.1 From *top* to *bottom*, algorithms 1 to 3: drift correction model, adaptive table for NO_x estimation and adaptive NO_x model

6.2 Fast Estimation of λ^{-1}

The estimation of λ^{-1} is the core of this section, where a complete methodology based on the drift correction model developed in Sect. 5.2.1 is proposed (Guardiola et al. 2013c). The original problem is modified in order to consider the model and sensor uncertainties as well as the actual variation of the bias $\theta_{\lambda^{-1}}$ with respect to the operating point and time as remarked in (4.3). In addition, the variable λ^{-1}

is a relevant input for the NO_x model, and its observation by KF methods can be integrated in an adaptive NO_x model. First, the problem set-up and the methodology are explained.

6.2.1 Problem Set-up and Methodology

The goal is providing an estimation $\hat{x}_{\lambda^{-1}}$ of the actual value of λ^{-1} combining the information provided by the sensor $z_{\lambda^{-1}}$ and the model $x_{\lambda^{-1}}$. The basic idea is keeping the model dynamics while correcting the model bias (and its drift) profiting the sensor steady performance.

Figure 6.2 shows an schematic representation of the proposed structure. All calculations and equations are set-up in discrete form where represents the sampling period, which corresponds to 20 ms for the results shown in this work. Main inputs are the injected fuel mass m_f, the air mass flow m_a, the speed n and the sensor signal $z_{\lambda^{-1}}$.

Because of the bias strong dependence with the operating point conditions, an adaptive 2D look-up table $\mathbf{\Theta}_{\lambda^{-1},\mathbf{k}} \in \mathbb{R}^{n_r \times n_c}$ updated at instant k is used for accounting with this variation depending on the engine speed and load (see discussion in Sect. 4.3.1 about engine operating point representation). Hence the current bias at the instant k may be represented as

$$\theta_{\lambda^{-1}}(k) = \mathbf{\Theta}_{\lambda^{-1},\mathbf{k}}(n(k), m_f(k)) + w_k \tag{6.1}$$

where a noise $w_k \in \mathbb{R}$ is added to the tabulated value coming from the table $\mathbf{\Theta}_{\lambda^{-1},\mathbf{k}}$ at instant k for considering modelling errors, and for dealing with the system drift.

Through the combination of the sensor and the model information (conveniently delayed for being comparable), a KF is used for tracking the value of the bias and of the filtered value of λ^{-1}, thus providing the estimates $\hat{\theta}_{\lambda^{-1}}$ and $\hat{x}_{\lambda^{-1}f}$. A set of 'freezing' conditions adds robustness to the algorithm stopping the integration when the sensor signals or the model are not reliable.

In order to cope with the system drift, the look-up table $\mathbf{\Theta}_{\lambda^{-1}}$ is updated on the basis of the adaptive SSKF algorithm that uses the estimate $\hat{\theta}_{\lambda^{-1}}$

$$\mathbf{\Theta}_{\lambda^{-1},\mathbf{k}} = f(\mathbf{\Theta}_{\lambda^{-1},\mathbf{k}}, \hat{\theta}_{\lambda^{-1}}(k)) \tag{6.2}$$

Finally, the estimation $\hat{x}_{\lambda^{-1}}$ is built up from the current model output $x_{\lambda^{-1}}$ and the tabulated value of the bias (see (6.5)).

6.2.2 Robustness Against Signals Uncertainties

Three main circumstances affect the correct algorithm performance:

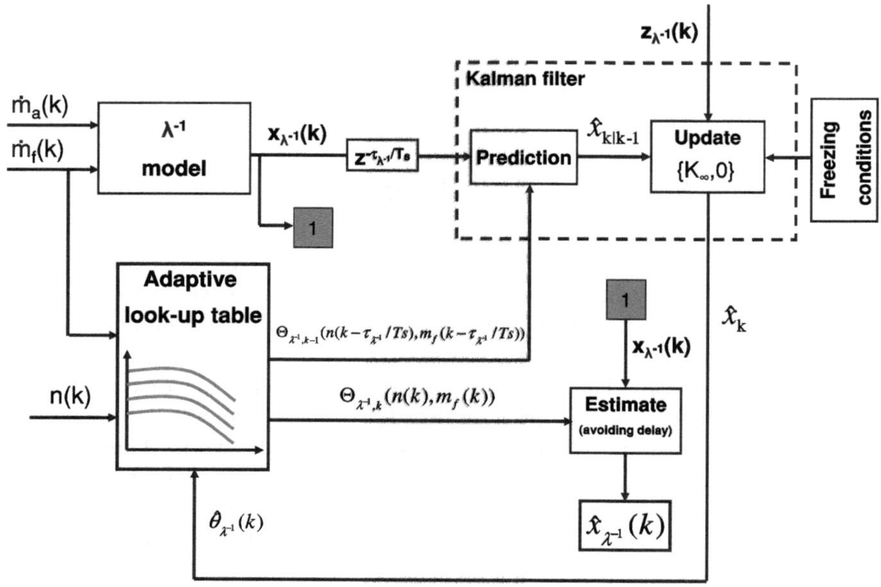

Fig. 6.2 Schematic view of the proposed procedure for λ^{-1} estimation

- Sensor saturation: NO_X sensors exhibit saturation problems for high O_2 partial pressures (low λ^{-1}). Figure 6.3 shows the saturation point for the tested NO_X sensor, where two units of this model are located downstream of the turbo and downstream of the engine here equipped with an AT line. If the KF is active, the integration would be incorrect in these situations as far as $z_{\lambda^{-1}}$ is not reliable.
- Sensor and model uncertainties during sharp dynamic transients: The sensor behaviour, defined by $a_{\lambda^{-1}}$ and $\tau_{\lambda^{-1}}$, varies with time, boundary conditions, system conditions and because of the ageing and unit-to-unit dispersion. For avoiding wrong integrations, these must be considered. Furthermore, the sensor model considered in (3.3) could not be complex enough for accounting with the actual sensor performance. An incorrect sensor model leads to peak errors, especially when a sharp transient occurs.
- Spurious measurements: Signals involved can present outliers or errors that could be fatal for the algorithm.

When these occur, the integration must be stopped. A set of deactivation IF- THEN or 'freezing' rules have been programmed for that. If any of the following conditions is true, the updating of the KF is null; i.e. $e_k = 0$:

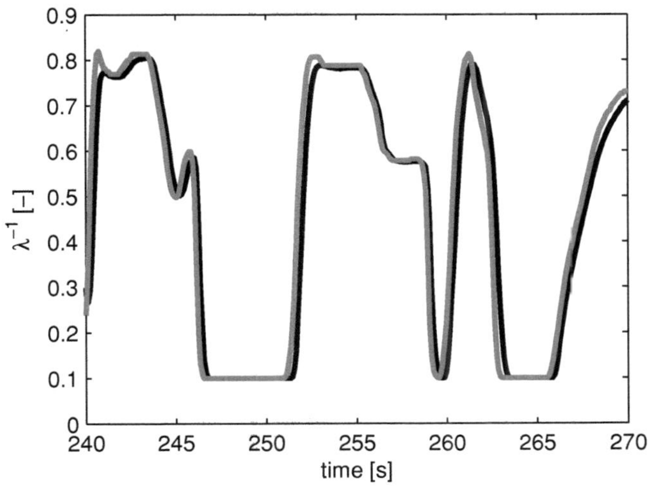

Fig. 6.3 $z_{\lambda^{-1}}$ measurements from two NO_x sensors, the first located downstream of the turbine (*gray*) and the second downstream of the AT line (*black*) in an engine equipped with DOC, SCR and DPF (see the third configuration of Fig. 5.1). Both sensors are the same commercial model and present a saturation in approximately $\lambda^{-1} = 0.1$

$$e_k \frac{1 - z^{-1}}{} > F_1 \tag{6.3a}$$

$$x_{\lambda^{-1}} \frac{1 - z^{-1}}{} > F_2 \tag{6.3b}$$

$$z_{\lambda^{-1}} \frac{1 - z^{-1}}{} > F_3 \tag{6.3c}$$

$$z_{\lambda^{-1}} < F_4 \tag{6.3d}$$

$$z_{\lambda^{-1}} > F_5 \tag{6.3e}$$

$$\hat{x}_{\lambda^{-1}} < F_6 \tag{6.3f}$$

$$\hat{x}_{\lambda^{-1}} > F_7 \tag{6.3g}$$

where (6.3a)–(6.3c) deactivate the filter in sharp dynamic transients or in the case of spurious measurements; (6.3d) if sensor saturation or outliers appear; (6.3e) if measurement outliers are detected; finally (6.3f) and (6.3g) are included for providing robustness to the estimation. These conditions make the table update highly insensitive to sensor model errors, sensor output errors and model output errors.

The freezing conditions allows to set a faster integration, i.e. a higher Kalman gain, as integrations at points with high expected errors on $\hat{\theta}_{\lambda^{-1}}$ are avoided. The definition of thresholds needs to be done accordingly with the level of uncertainty in the sensor knowledge, signals noise level, model reliability and assumptions concerning the dynamic characteristics of the λ^{-1} evolution.

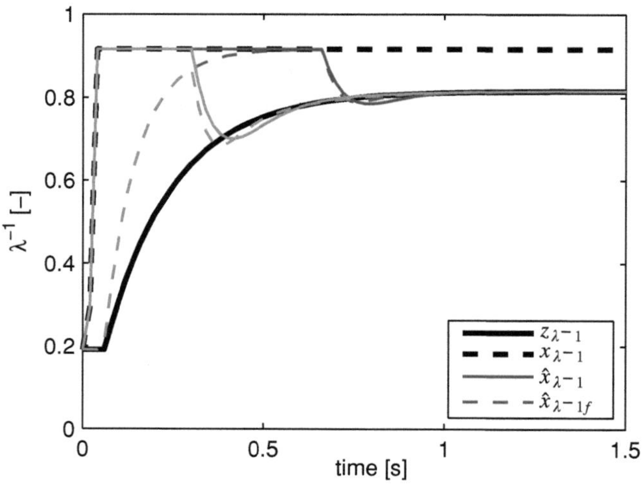

Fig. 6.4 $\hat{x}_{\lambda^{-1}}$ for two different conditions of freezing for an injection step from 15 to 30 mg/str at 2,250 rpm. *Thin solid lines* represents $\hat{x}_{\lambda^{-1}}$, lighter to darker *gray* is less to more severe freezing condition. $\hat{x}_{\lambda^{-1}f}$ models linked with the estimate $\hat{x}_{\lambda^{-1}}$ are also included with *dashed lines*

Figure 6.4 compares $\hat{x}_{\lambda^{-1}}$ obtained with two different tunings of the thresholds and for one sharp injection transient, where λ^{-1} value is drastically varied and the bias is affected by the operating conditions (it varies from 0 at the beginning of the test to 0.1 after the step). In all cases, a significant error in the sensor model is assumed ($a'_{\lambda^{-1}} \neq a_{\lambda^{-1}}$ and $\tau'_{\lambda^{-1}} \neq \tau_{\lambda^{-1}}$). It can also be noticed that including freezing mitigates the overshoot in the correction; although a conservative tuning will result to completely deactivating the integration of the filter, and then no bias integration would be made during the engine operation. The Monte Carlo calibration method presented in Payri et al. (2012) might be used for tuning the thresholds F_1 to F_7.

6.2.3 An Adaptive Look-up Table for Modelling the Drift

An adaptive look-up table (see Fig. 6.2) is used for coping with the bias dependence with the operative conditions. The look-up table interpolation principle is based on a 2D bilinear interpolation. At every iteration, the initial estimation of the bias is interpolated from the table $\mathbf{\Theta}_{\lambda^{-1}}$ conveniently updated

$$\hat{\theta}_{\lambda^{-1}}(k|k-1) = \mathbf{\Theta}_{\lambda^{-1},\mathbf{k-1}}(n(k-\tau/), m_f(k-\tau/)) \tag{6.4}$$

where the table scheduling inputs n and m_f are conveniently delayed in order to consider the sensor delay. The KF provides an updated value of the bias (at the

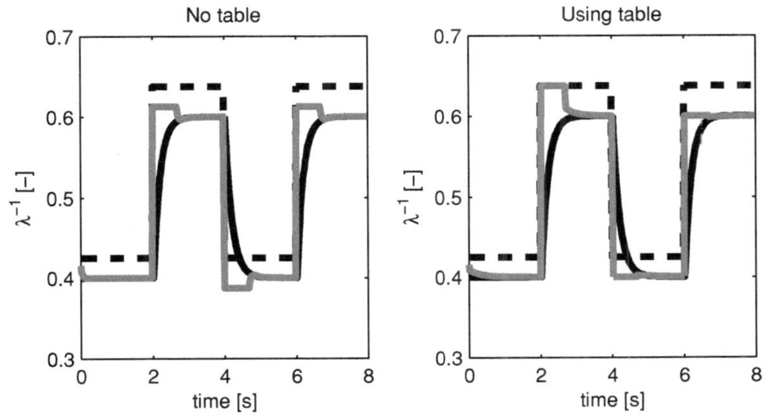

Fig. 6.5 $\hat{x}_{\lambda^{-1}}$ without table (*left*) and using an adaptive table (*right*) changing the operating point condition of the engine repeating two times a cycle by simulating. *Black line* is $z_{\lambda^{-1}}$, *dashed black line* is $x_{\lambda^{-1}}$ and *gray line* is $\hat{x}_{\lambda^{-1}}$. Deactivation freezing rules are applied

delayed input conditions) that is used for updating the table with a given learning method. Chapter 5 have made a review and developed two methods based on the EKF for updating look-up tables. For this formulation, the SSKF method presents the advantage that the table can be considered as an independent block, and then the converged K_∞ for the drift correction is still applicable. In this section, the results by using the SSKF are presented (although for a robust implementation, the SKF is advised).

Once the table has been adapted, an updated value of the bias is obtained for the current operative conditions. This is used for building the final estimation

$$\hat{x}_{\lambda^{-1}}(k) = x_{\lambda^{-1}}(k) + \mathbf{\Theta}_{\lambda^{-1},\mathbf{k}}(n(k), m_f(k)) \tag{6.5}$$

where the current values of the inputs and the model are used at the instant k, in order to provide the more updated information and avoiding the causality problem related with the sensor delay when the table is not used.

For illustrating the look-up table performance, Fig. 6.5 shows the simulation of λ^{-1} for a repetition of the same profile. $\tau_{\lambda^{-1}}$ is known and null and $a_{\lambda^{-1}}$ presents a certain error. The top plot represents the evolution if the adaptive table is not used. In the case of using the adaptive table (right plot), first part of the cycle serves for the bias identification (initially all elements of the table are set to zero), and in the second part of the cycle the estimation is significantly improved because of the stored values. Note that during transients, the error integration starts slightly after the step because of the freezing rules.

Different tables may also be used in case of using different combustion modes (split injection, exhaust system regeneration) for coping with the different trend of the bias on the m_f estimate when the injection settings change. Each table is updated only when the mode is activated and can add more accuracy to the estimate although incrementing the programming complexity.

6.2.4 Experimental Results

In order to provide a reference signal for assessing the algorithm, injection steps are performed in the engine and $\hat{x}_{\lambda^{-1}}$ calculated. Since the injection is a fast acting control variable, the actual λ^{-1} response is judged to be instantaneous and no more than one engine cycle delay is expected; being able to generate a reference for comparison. Furthermore, this signal is available online, making that the algorithm may be tuned and proved without a specific test rig beyond the engine and the ECU itself.

The Sensor model identification was discussed in Chap. 3, hence a delayed first order filter ((see (3.3)) with parameters $a_{\lambda^{-1}} = 0.92$ and $\tau_{\lambda^{-1}} = 0.5$ s with a sampling frequency of 50 Hz (= 0.02 s) is used. K_∞ is precalculated offline for all cases as the KF is steady-state. Figure 6.6 shows the results for a set of different injection steps for a constant engine speed $n = 1{,}500$ rpm. For that figure, the ZrO_2 sensor has been moved downstream of the AT systems and a significant bias, 5 % of its own value, has been added for exaggerating the correction; i.e $x_{\lambda^{-1}}^{drift} = 1.05 x_{\lambda^{-1}}$. Furthermore, $x_{\lambda^{-1}}$ presents a significant variable bias depending on the operating point with respect to $z_{\lambda^{-1}}$. The drift correction algorithm is applied to the estimation of λ^{-1}. Freezing conditions are relaxed because the sensor is well modelled. The left plot shows an example where the integration is quite slow; the right plot shows the opposite situation where the integration is quite fast, provoking some peaks but with a lower settling time. Here, no model or table is used for modelling $\theta_{\lambda^{-1}}$. This makes that after every step, $\theta_{\lambda^{-1}}$ shall be integrated for coping with its variation.

The use of an adaptive table $\boldsymbol{\Theta}_{\lambda^{-1}}$ smoothes the bias integration when the operating point changes once the table has been learnt. Figure 6.7 shows the same situation of Fig. 6.6 but now using a look-up table with the grid

$$r = [750 \quad 1{,}000 : 500 : 5{,}000] \tag{6.6a}$$

$$c = [0 : 5 : 30 \quad 34 : 4 : 50 \quad 55 : 5 : 80] \tag{6.6b}$$

Figure 6.8 shows final results for three different injection steps of the previous cycle once that the bias has been perfectly learnt. $\hat{x}_{\lambda^{-1}}$ keeps fast dynamics but converges to the steady-state value of $x_{\lambda^{-1}}$, in contrast to the biased model and the slow sensor.

Finally, the algorithm is proven with the NEDC B1 cycle and is shown in Fig. 6.9 for a window of the cycle. The described conditions and data for previous tests are still valid here. $\hat{x}_{\lambda^{-1}}$ keeps the model dynamics but evolves correcting the drift and relying on the steady-state value of the sensor. This test is representative of real driving conditions and the procedure demonstrates its feasibility for being used in commercial vehicles.

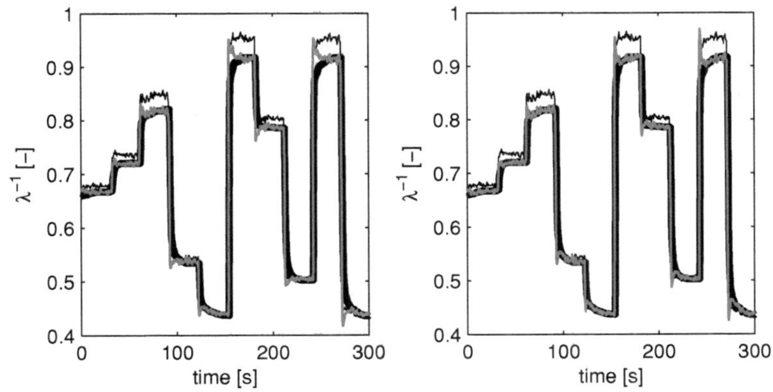

Fig. 6.6 $\hat{x}_{\lambda^{-1}}$ performing injection steps at 1,500 rpm without using a look-up table. *Left plot* Slow correction, with $\sigma_w^2/\sigma_v^2 = 10^{-5}$. *Right plot* Fast correction, with $\sigma_w^2/\sigma_v^2 = 10^{-4}$. Legend: *Thick black* is $z_{\lambda^{-1}}$, *thin black*is $x_{\lambda^{-1}}$ and *gray* is $\hat{x}_{\lambda^{-1}}$

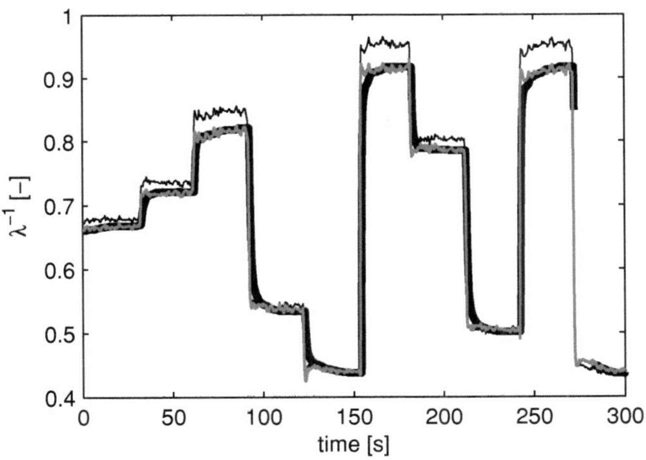

Fig. 6.7 $\hat{x}_{\lambda^{-1}}$ performing injection steps at 1,500 rpm and using an adaptive look-up table for modelling $\theta_{\lambda^{-1}}$. Legend: *Thick black* is the sensor $z_{\lambda^{-1}}$, *thin black* is the model $x_{\lambda^{-1}}$ and *gray* is $\hat{x}_{\lambda^{-1}}$. Values for tuning the KF are $\sigma_w^2/\sigma_v^2 = 1/80$

6.3 Fast Estimation of NO_x

The attention is turned out to the problem of the online NO_x estimation. With an algorithm like B, an adaptive NO_x map can be used in order to infer NO_x as function of only m_f and n as in (4.8), which is justified if the transitions are slow enough for utilising a quasi-static approach (plus a filter for the sensor dynamics). An algorithm C permits to update NO_x model parameters and look-up tables. Finally, and for both

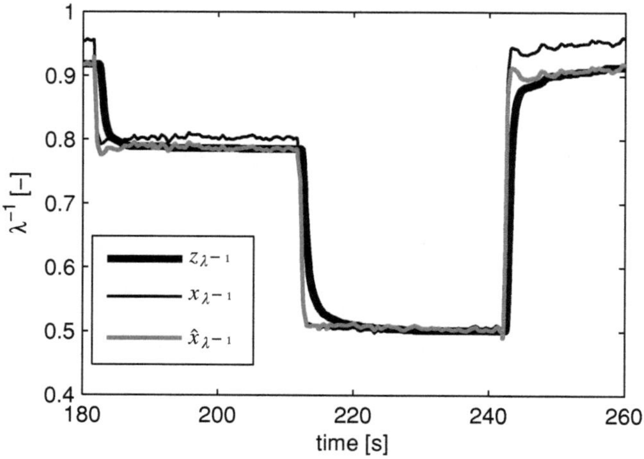

Fig. 6.8 $\hat{x}_{\lambda^{-1}}$ using table and freezing conditions performing different injection steps at 1,500 rpm

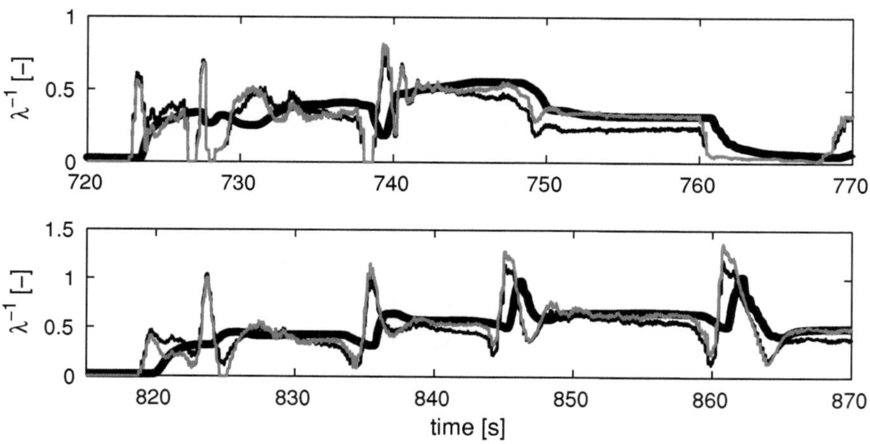

Fig. 6.9 *Top* and *bottom* $\hat{x}_{\lambda^{-1}}$ using the table and applying freezing conditions to the NEDC B01 cycle. $z_{\lambda^{-1}}$ is the sensor output (*thick black*), $x_{\lambda^{-1}}$ is the model output (*thin black*) and $\hat{x}_{\lambda^{-1}}$ is the observation (*gray*). The drift is corrected while the dynamics of the observation $\hat{x}_{\lambda^{-1}}$ are maintained

designs, the algorithm A and the drift correction algorithm could be used in order to track the model bias.

This section also completes the comparison of the different methods for updating look-up tables already presented in Chap. 5 by testing the cycle SDMP B5 and the algorithm B. Afterwards, the adaptive tuning of the RT NO_x model is proposed. On the basis of the presented results, the estimation of the actual NO_x is proposed.

6.3.1 Online Updating of Look-up Tables for Modelling NO_x

In a first approach, NO_x is modelled by using a 2D adaptive look-up table function of n and m_f, whose parameters are estimated with the updating methods (Guardiola et al. 2013a, b). The state-space model (5.33) is used for designing an observer like the algorithm B presented in Fig. 6.1.

The model parameters are estimated online while the engine is running without any special calibration procedure or test rig, beyond the sensor measurements. The first time that the engine is running, parameters evolve, and when the engine switches off, the stored parameters can be used for predicting NO_x. When the engine is running again, the parameters keep evolving for correcting drift and slowly varying effects. Furthermore, the observer built for updating the map can be profited for having an actual NO_x estimation, i.e. avoiding the filtering and the delay of the sensor. The SKF updating method is appropriate for online usage because of the light computational burden and the estimation capabilities.

Moreover, this configuration might be applied to estimate local models for defining the NO_x emissions as function of u_{soi} and u_{egr}, and for different engine operating points (Desantes et al. 2011; Arsie et al. 2012). These maps are useful for the joint air and fuel path control (Tschanz et al. 2012).

With respect to the dimensions of the maps used for the adaptive estimation of NO_x and according to the author's experience, usual dimensions for look-up tables in production engines are between 200 and 400 parameters. The selection of the appropriate dimensions is a trade-off between the accuracy and the computational resources. An excessive density can compromise the table convergence, but a higher density of scheduling points can improve the accuracy of the solution.

Figure 6.10 shows the nominal $NO_{x,0}$ map used for the NO_x model in Chap. 4. This map was filled with an intensive DOE comprising 363 steady-state tests. The NO_x sensitivity depends on the engine area, with a clear difference between the EGR area, transition and the areas without EGR. Note that all elements of $NO_{x,0}$ are filled in order to avoid numerical problems, indeed areas that are not reached for the engine (out of the full load line).

The grid used in this work for the observation is defined as follows

$$r = [750 \quad 1{,}000 \quad 1{,}250 \quad 1{,}500 \quad 1{,}750 \quad 2{,}000 : 500 : 4{,}500] \tag{6.7a}$$

$$c = [0 : 5 : 30 \quad 34 : 4 : 50 \quad 55 : 5 : 80] \tag{6.7b}$$

where r represents the grid for n (rpm) and c stands for m_f (mg/str).

6.3.1.1 Comparison of the Updating Methods

The SDMP B5 cycle is used for comparing the algorithms with real engine data. $\mathbf{\Theta}_{NO_x,0}$ is the null matrix and is updated with the cycle. The state-space model (5.33)

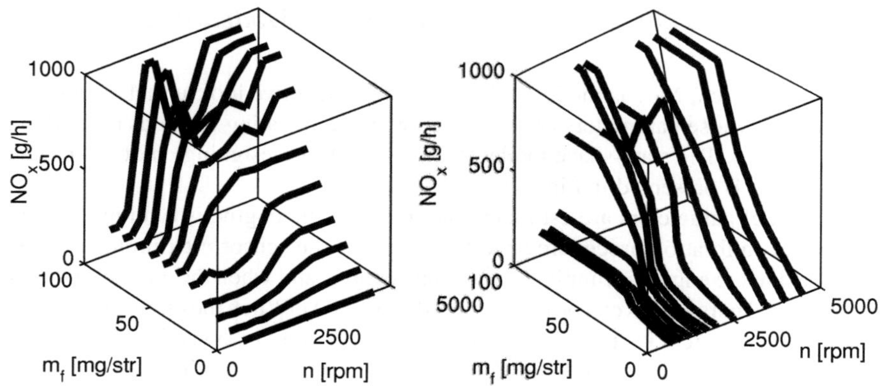

Fig. 6.10 NO_x map plotted in the n and m_f directions. *Left plot* represents the NO_x emissions at iso-m_f values, while *right plot* shows the NO_x emissions at iso-n

is built with the following numerical values

$$a_{NO_x} = 0.96; \quad \tau_{NO_x} = 0.75\,\text{s} \tag{6.8}$$

while the sampling frequency is 50 Hz (=20 ms). Sensor parameters come from the methods presented in Chap. 3 by performing SOI steps. The state-space model output z_{NO_x} is given by the NO_x sensor, while the inputs are

$$u_{1,k} = z^{-\tau_{NO_x}/}\, n(k) \tag{6.9a}$$

$$u_{2,k} = z^{-\tau_{NO_x}/}\, m_f(k) \tag{6.9b}$$

and the state vector x is

$$x_k = [vec(\boldsymbol{\Theta}_{NO_x}) \;\; x_{NO_xf}]_k^T \tag{6.10}$$

The filter is tuned with the values

$$\sigma_v^2 = 25^2; \quad \sigma_w^2 = 15^2; \quad \sigma_f^2 = 50^2 \tag{6.11}$$

and a test subset of $t = 400\,\text{s}$ is used for updating the adaptive map and the look-up table $\boldsymbol{\Theta}_{NO_x,400}$ is used for predicting NO_x in the whole cycle plus considering the first order filter that models the sensor response.

$$y_{NO_x} = \frac{1 - a_{NO_x}}{1 - a_{NO_x} z^{-1}} \boldsymbol{\Theta}_{NO_x,400}(u_{1,k}, u_{2,k}) \tag{6.12}$$

The KF, SKF and SSKF methods are used for updating the model and the results can be seen in Fig. 6.11. All three methods behave well and are capable of predicting

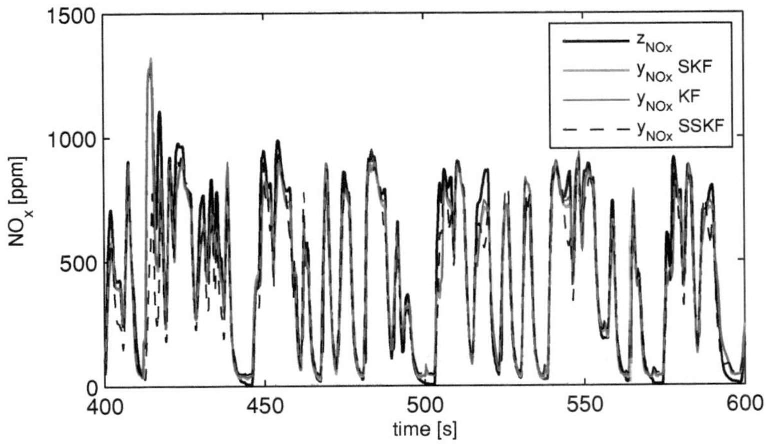

Fig. 6.11 NO_x prediction for the three methods by performing the SDMP B5 cycle. The maps are updated during 400 s, i.e. a subset of 20,000 measurements of z_{NO_x}. y_{NO_x} is simulated with the updated tables at 400 s $\mathbf{\Theta}_{NO_x,400}$. The sensor measurement z_{NO_x} is provided for comparison

NO_x emissions. Dynamics assumptions are also good enough for having a good fitting in the SDMP. Note that the SSKF results are also acceptable as the covariance tracking is not relevant in this cycle as it is in slow varying tests. The fitting should be optimized by independently tuning the filter for each method.

It is worth comparing the convergence of the methods and for this, the SDMP cycle is also used as benchmark for testing the absolute mean error when the updating time is varied monotonically around the cycle

$$e_M(k) = \frac{\sum_{i=1}^{n_y} \left(z_{NO_x}(i) - y_{NO_x}(i)|_{t=k}\right)}{k} \tag{6.13a}$$

where t is the time that the table has been updated and y_{NO_x} is calculated by (6.12). The results are shown in Fig. 6.12. Focusing the attention in the lines generated with the filter calibration Cal (6.11), e_M tends to be lower as t grows at least during first iterations when the system knowledge is poor. The horizontal line shows the benchmark model error, which is constant as the model calibration is fixed. It is clear how all the three methods converge to an error similar to that of the benchmark model, but with the advantage that the drift and ageing are considered by the online versions but not by the model. The KF and SKF behaviour is quite similar as expected. On the other hand, Fig. 6.12 also shows that the KF and the SKF are faster than the SSKF and this is due to the filter tuning. Anyway, there exists a trade-off between the convergence speed and the estimation robustness, e.g. the KF and SKF have a significant oscillation with respect to the SSKF around the benchmark model line. This is a matter of the noise tuning and the results are also generated by using a lower gain observer with the calibration set

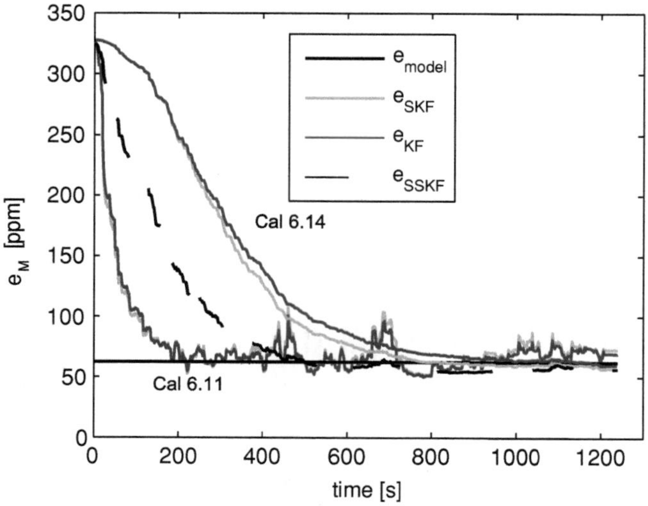

Fig. 6.12 Evolution of the mean of the absolute error e_M for the different methods in the SDMP B5 cycle and for two different filter calibrations. SSKF *dashed line* is calculated by using the filter calibration (6.11). *Lines left* to the SSKF one using calibration (6.11). During first iterations, the error is high, as far as the table has no initial knowledge, but once that the time is around 400 s, the error is minimum, although the noise transmission is evident. *Lines right* to the SSKF one using calibration (6.14), noise transmission is reduced but the convergence speed as well. The chosen filter tuning must be a trade-off between the robustness and convergence speed

$$\sigma_v^2 = 25^2;\ \sigma_w^2 = 7^2;\ \sigma_f^2 = 50^2 \tag{6.14}$$

for the KF and the SKF. The results are shown in lines generated by Cal (6.14) in Fig. 6.12. With this configuration, the convergence speed is lower but the noise transmission and overfitting is avoided when t is large. This goes in the direction of the robustness and the optimised calibration data set must solve this trade-off, considering uncertainties in the sensor behavior knowledge or in the model quality and data-set quality, among others. Strategies such as the freezing applied to the λ^{-1} estimation may also be applied to this problem.

With respect to the global observability, variances of the 4 table parameters are compared for both the KF and SKF using the calibration (6.14). Figure 6.13 shows the results. The top plot shows an element that is never observed during first 200 s and due to this, the variance increases monotonically and x_{126} is still null in $t = 200$. Second and third elements represent two elements that are active at some instants. This is clearly seen when the variance decreases, while when the elements are not active, the variance increases again monotonically. Note that in a production system application, that will run for the life time of a vehicle, the elements of the estimation error covariance matrix need to be limited so that they do not grow too much and cause numerical problems, see e.g. Höckerdal et al. (2011) who propose a saturation

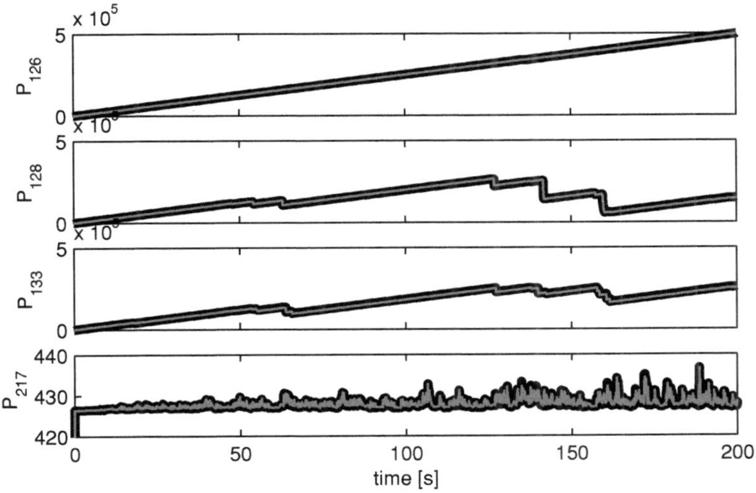

Fig. 6.13 Variance analysis by updating a NO_x map in the cycle SDMP B5. From *top* to *bottom*, variances of elements 126, 126, 133 and 217 are plotted. *Black line* is the KF and *gray line* is the SKF

for avoiding too large variances. The bottom plot shows the variance related to $x_{NO_x f}$ that is fairly constant and is a proof of the system global observability.

The variances by using the KF and the SKF behave similar and differences can only be found when zooming in the Fig. 6.13. Therefore, the SKF method performs quite similar to the standard KF but with a much lower computational burden. Due to the SDMP presents sharp transients, the SSKF method also performs quite well, and proves that this method can be useful when the considered data set fulfils some conditions in order to ensure robustness. To the nature of the application considered here, the SKF is the best solution for online updating of maps.

6.3.1.2 Adaptive Maps for Predicting NO_x

The SDMP is a test with sharp variations on the operating point conditions but the homologation cycles, such as the NEDC, are much slower. This could compromise the global observability since higher levels of excitation are beneficial for updating. As shown in Fig. 4.3, the air path dynamics are fast enough to follow the load variations in the NEDC and then $\dot{m}_a$ and p_{boost} are able to track their references. In such case, the model (4.8) is valid and the table $\mathbf{\Theta}_{NO_x}$ might directly replace the nominal $NO_{x,0}$ table in the RT NO_x model ((see (4.22)). Despite $NO_{x,0}$ could be used for initializing $\mathbf{\Theta}_{NO_x}$, here the null matrix is the initial map, which is a worst-case condition.

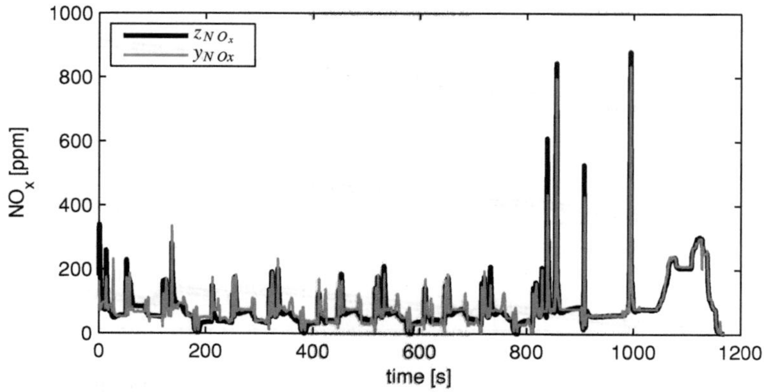

Fig. 6.14 Adaptive NO_x estimation in the NEDC A01, where it is estimated with a table that is updated during the whole cycle and then used for simulating y_{NO_x}. z_{NO_x} is provided for comparison. The estimation keeps the dynamic properties of z_{NO_x} but after updating, the drift in Fig. 4.19 is cancelled

The NEDC A1 cycle is run for updating the map. Figure 4.19 already showed the RT NO_x model output when comparing with z_{NO_x} in that cycle, where a clear drift exists. After using an adaptive map and the SKF method with the calibration (6.14), the results are shown in Fig. 6.14. There, y_{NO_x} is calculated on the basis of a table $\boldsymbol{\Theta}_{NO_x}$ which has been updated during the whole cycle. After comparing with z_{NO_x} and x_{NO_x} coming from the NO_x model (see Fig. 4.19), the adaptive map approach gets a good NO_x estimation which is slightly better than the one of the NO_x benchmark model.

The SKF could be used for the online adaptation and/or calibration of complex models, where a large number of maps and parameters must be updated. Anyway, a deep study is required for ensuring the observability, convergence and robustness properties and for getting a computationally efficient learning structure. A big number of parameters and maps should be updated and the problem is non-convex. Admittedly the computational issue is critical and there the SKF can be an effective solution.

In the case of a fast cycle, i.e. the SDMP, the EGR valve is closed the most of the time. Therefore, the states are function of n and m_f for the engine settings corresponding to a closed EGR valve

$$x = f(n, m_f, m_a^*(n, m_f), p_b^*(n, m_f))_{u_{egr}=100} = f(n, m_f)_{u_{egr}=100} \tag{6.15}$$

In the case of the NEDC or CADC cycles, the transients are much slower and the engine works closer to the nominal settings. In such case, the states are also function of n and m_f but for the nominal operation of the EGR-VGT controller. Possible states can now be approximated with

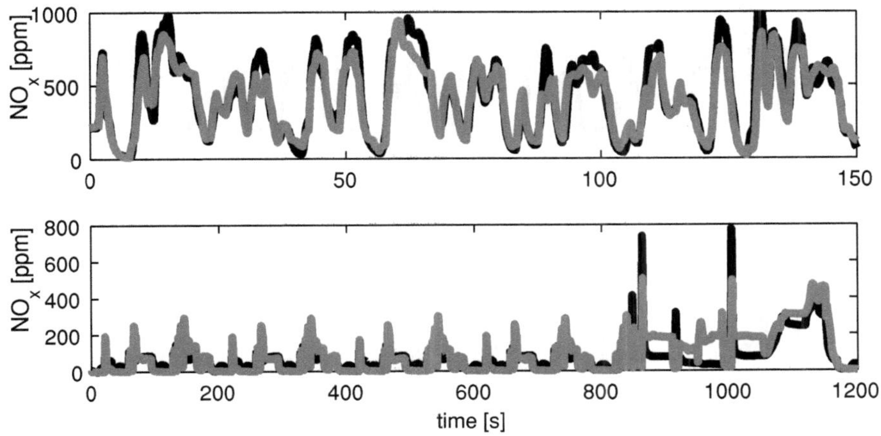

Fig. 6.15 Offline NO_x prediction y_{NO_x} (*gray*) against NO_x sensor (*black*) by using a table $\boldsymbol{\Theta}_{NO_x}$ updated online in the SDMP B5 cycle. *Top plot* results in the SDMP B6 cycle. *Bottom plot* results in the NEDC B1 cycle

$$x = f(n, m_f, m_a^*(n, m_f), p_b^*(n, m_f))_0 = f(n, m_f)_0 \tag{6.16}$$

This point links with the discussion of Sect. 4.3.1 about how to represent the engine conditions. Figure 6.15 shows the predicted NO_x emissions in the cycles SDMP B6 (top plot) and NEDC B1 (bottom plot) after using a map $\boldsymbol{\Theta}_{NO_x}$ that has been updated in the cycle SDMP B5. The table output is filtered and delayed by a sensor model with the parameters $a_{NO_x} = 0.96$ and $\tau_{NO_x} = 0.75$s. The ability to reproduce NO_x on the SDMP B6 is clear while a slight drift exists in the NEDC B01. Anyway, the NO_x estimation during the urban part is not as bad as in the highway part, where NO_x is overestimated as expected ($\boldsymbol{\Theta}_{NO_x}$ has been fitted with nearly zero EGR conditions).

However, if updating $\boldsymbol{\Theta}_{NO_x}$ with the NEDC B1 cycle and then using the results for predicting the SDMP B6, shown in top plot of Fig. 6.16, the results are quite bad as a big area of the engine map is not excited and the covered areas present lower NO_x emissions because of the EGR actuation. The prediction using $\boldsymbol{\Theta}_{NO_x}$ in the CADC B2 urban cycle is quite better, as the covered areas are similar.

6.3.2 Online Updating of the NO_x Model

In Chap. 4, a NO_x model was developed based on a set of look-up tables for representing nominal engine conditions with several corrections based on the estimated intake oxygen, thermal loading and others as in (4.22). In this section, adaptive capabilities are proved in the NO_x model (Schilling et al. 2006), and the new state-space vector X^w might be augmented as follows

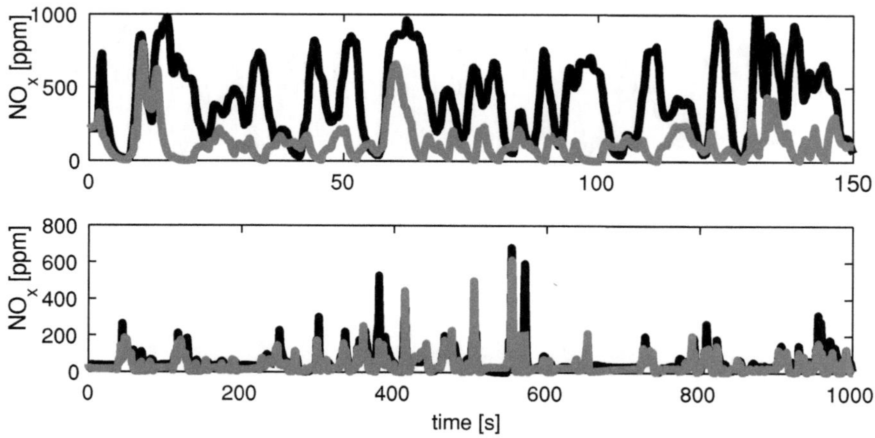

Fig. 6.16 Offline NO_x prediction y_{NO_x} (*gray*) against NO_x sensor (*black*) by using a table Θ_{NO_x} updated online in the NEDC B1 cycle. *Top plot* results in the SDMP B6 cycle. *Bottom plot* results in the CADC B2 cycle

$$X^w = \begin{bmatrix} x_{NO_x} & y_{NO_x} & x_{\lambda^{-1}} & y_{\lambda^{-1}} & vec(\Theta_{NO_x}) & \theta \end{bmatrix}^T \tag{6.17}$$

where besides x_{NO_x} and y_{NO_x}, the following states are included: $x_{\lambda^{-1}}$ and $y_{\lambda^{-1}}$ representing the actual λ^{-1} and the filtered λ^{-1} (for comparing with the λ^{-1} output from the sensor as was made in (6.12) for NO_x) respectively and used as inputs for the RT NO_x model, the nominal NO_x map Θ_{NO_x} and the model bias θ_{NO_x}.

On one side, the use of an observer for λ^{-1} if the NO_x sensor is available is completely justified due to the NO_x sensitivity to errors on mistaking the intake oxygen (Arrègle et al. 2008) (see Table 4.2). Furthermore, the error sources in the λ^{-1} model were already explained in Sect. 4.2 and an example of λ^{-1} observation is shown for the NEDC in Fig. 6.9 where the model drift is not negligible. Figure 6.17 precisely compares the model output y_{NO_x} when a drift on $x_{\lambda^{-1}}$ estimation occurs. The top plot shows $z_{\lambda^{-1}}$ by a ZrO_2 sensor, $x_{\lambda^{-1}}$ from (4.1) and the observation $\hat{x}_{\lambda^{-1}}$ by a KF, when a clear drift on $x_{\lambda^{-1}}$ exists.[1] If the drifted signal is used for inferring NO_x by the model, the output is also drifted from the sensor signal as shown in the bottom plot. However, if observing $\hat{x}_{\lambda^{-1}}$ by a KF, the NO_x output keeps the model accuracy. Therefore, the benefits of observing λ^{-1} are clear.

On the other side, model errors and ageing can be corrected online by tracking the bias in the same way as explained in Sect. 6.2 for λ^{-1}, or by observing the nominal $NO_{x,0}$ table for also improving the prediction capabilities of the model. The former is useful for actual NO_x prediction, while the latter is advisable for ensuring the model accuracy.

[1] The drift on $x_{\lambda^{-1}}$ is caused due to failures in the air mass flow or the injected fuel mass signals. This drift will affect to the EGR flow model and the look-up tables interpolation. In those cases, the sensor signal $z_{\lambda^{-1}}$ can be used for correcting errors on the air mass flow and/or injected fuel mass signals. In this section and for the simulation, the error is only considered in the $x_{\lambda^{-1}}$ model.

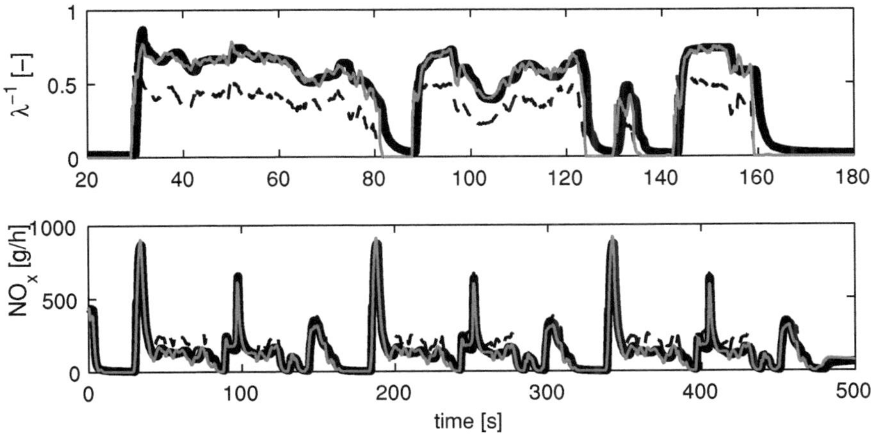

Fig. 6.17 λ^{-1} and NO_x estimations in a TRAN B ($T_{cool} = nom$, $\Delta u_{egr} = 0$ %, $\Delta p_{boost} = 0$). *Top plot* $z_{\lambda^{-1}}$ sensor signal (*black*), $x_{\lambda^{-1}}$ calculated by the model (*dashed thin black*) and $\hat{x}_{\lambda^{-1}}$ observed by a drift augmented model (*gray*). *Bottom plot* NO_x estimation by three ways, z_{NO_x} sensor signal (*black*), y_{NO_x} by the NO_x model using $x_{\lambda^{-1}}$ as input (*dashed thin black*) and y_{NO_x} by using an adaptive model with $\hat{x}_{\lambda^{-1}}$ as λ^{-1} input

6.3.2.1 A Learning Algorithm for Updating the NO_x Model

The state-space vector

$$X^w = \begin{bmatrix} x_{NO_x} & y_{NO_x} & x_{\lambda^{-1}} & y_{\lambda^{-1}} & vec(\mathbf{\Theta}_{NO_x}) \end{bmatrix}^T \tag{6.18}$$

is built for designing an algorithm like C (see Fig. 6.1) in order to adapt the nominal NO_x table ($NO_{x,0}$) to possible changes in the NO_x emissions, mainly due to the ageing. Nevertheless, this structure is useful not only for learning $NO_{x,0}$ where

$$\mathbf{\Theta}_{NO_x} = NO_{x,0} \tag{6.19}$$

but for fitting a table without initial knowledge. Depending on the model and sensor uncertainties, the strategy of the filter tuning can be conservative (slow corrections) or aggressive (fast corrections).

The new state-space model remains similar to that of (5.33), used for the observer B, but including NO_x model variations and corrections around the nominal values ((see (4.22)) in the output matrix H_k

$$H_k^w = [zeros(length(vec(\mathbf{\Theta}_{NO_x}))) \quad C_{mod}] \tag{6.20}$$

where

$$C_{mod} = e^{-k_{NO_x} \cdot (EGR\lambda^{-1} - EGR_0\lambda_0^{-1})} \cdot C_{NO_x} \cdot C_{dyn} \tag{6.21}$$

Other model maps and parameters, such as the η_v, might be updated by just applying an EKF similar to that applied to $NO_{x,0}$ (with the correspondent linearisation). If a full adaptive model is proposed, in a way that all model parameters are observed, then the learning structure would lack of robustness and it would be difficult to discriminate the observed error between all tables, curves and coefficients. An alternative strategy could be to observe a drift θ_{NO_x} and afterwards weighting the error between model the parameters by LS or other strategies. Other limitations are the computational and memory resources or the lack of awareness in the error sources, making inadvisable to design an strategy based on the together updating of all model parameters and tables (Polóni et al. 2012).

$\boldsymbol{\Theta}_{NO_x}$ is designed with the grid defined by (6.7a), (6.7b) and is updated with the SKF method. The RT NO_x model is simulated with the tests of the campaign B and the errors are computed. Due to the system ageing, the model presents a drift. The SKF method is also used for updating $NO_{x,0}$ coming from the NO_x model and thus cancelling the drift. The updated model is simulated in those cycles and the new errors are computed.

For the application, the NO_x sensor output z_{NO_x} is converted from ppm to g/h by using (4.15) and the updating method is run with the output in g/h in order to directly modify $NO_{x,0}$ table.

The noise calibration is adjusted by the trial and error method and is not guaranteed to be the optimised one

$$\sigma_v^2 = 1;\ \sigma_w^2 = 0.01;\ \sigma_f^2 = 1 \tag{6.22}$$

The tests in the test campaign B are run by maintaining the calibration (6.22), and the offline NO_x prediction is calculated just after finishing the cycle by using the updated $NO_{x,0}$ in the NO_x model. The computed errors are compared with the ones of the model. The cycles used for the updating are the B1 to B6, totalling 6 tests. In these simulations, the initial matrix is $NO_{x,0}$ fitted for the NO_x model in Chap. 4. Figure 6.18 shows the results on 4 different cycles, where the original and updated y_{NO_x} are compared with the sensor signal z_{NO_x}. y_{NO_x} with the original calibration presents a bias with respect to z_{NO_x} that is corrected after running the cycles with the SKF.

Table 6.1 compares the absolute mean errors by using the original model output and the predictions of the updated model

$$e = mean(\sum_{k=1}^{k=l_{end}} z_{NO_x}(k) - y_{NO_x}(k)) \tag{6.23}$$

where l_{end} defines the number of samples taken for the considered cycle. The errors can be minimised achieving quite better results by using the learning structure. These results validate the SKF method and the capabilities for updating models.

With respect to the filter tuning, it has not been optimised for obtaining the results, and a trial-and-error method has been followed. Further improvements are in the line of optimising the tuning by improving the sensor model, e.g. including a variable

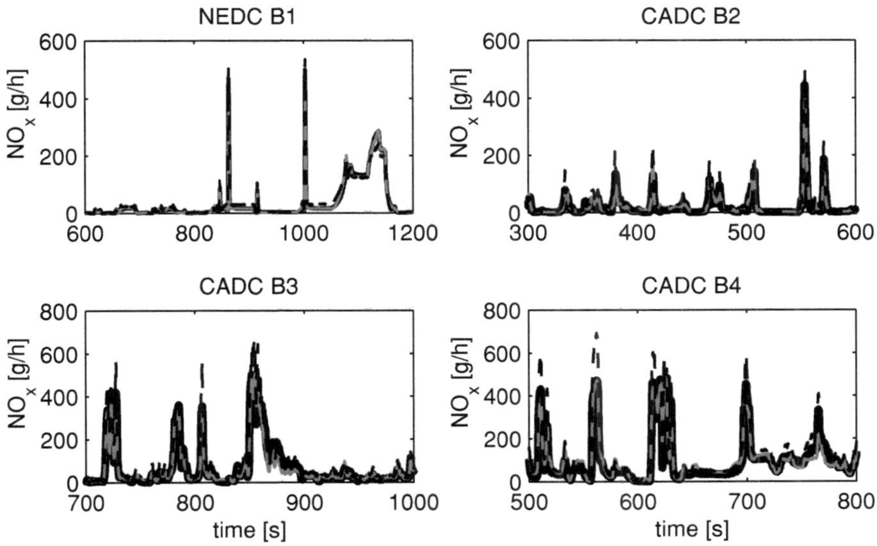

Fig. 6.18 Comparison of the NO_x estimation by three different sources in 4 different cycles: sensor signal z_{NO_x} (*black*), y_{NO_x} by using the RT NO_x model (*dashed thin black*) and y_{NO_x} computed by updating the $NO_{x,0}$ table with the SKF method (*gray*). *Left* to *right* and *top* to *bottom* NEDC B1, CADC B2, CADC B3 and CADC B4 cycles

Table 6.1 Mean absolute errors (g/h) in the NO_x model output for different tests, and comparing the error of the original model (model column) with the ones of the updated model (updated column)

Cycle	Model error [g/h]	Model error after updating [g/h]
NEDC B1	6.14	3.08
CADC urban B2	6.40	3.75
CADC rural B3	14.81	10.46
CADC highway B4	36.23	22.51
SDMP B5	49.46	49.13
SDMP B6	46.34	43.64

The cycles are run one time and the matrix $NO_{x,0}$ is updated online

delay, or augmenting the state vector with other model parameters. Anyway, the obtained results show the capabilities of the method for its online implementation in order to re-tune models and fit tables and parameters.

6.3.3 Online Observation of the Actual NO_x

The actual NO_x can be predicted by utilizing the state-space models and tables properly. For that, the state vector is augmented with an extra-state for representing the actual NO_x, i.e. NO_x emissions before sensor filtering and correcting the delay.

Estimating the actual NO_x is advisable for RT purposes, such as on-board NO_x control or diagnosis.

One of the important problems when of working on RT systems is the causality of the solution, as far as sensors usually measure in a delayed basis, mainly due to the transport and hardware delays. In this work, two possibilities are described for solving this issue: tracking the bias and the use of a model learning structure. Both possibilities are similar, and are based on the ageing cancellation by means of adaptive filtering.

6.3.3.1 Actual NO_x by Drift Correction Models

The following state-space vector is built

$$X^w = \left[\theta_{NO_x} \quad x_{NO_x} \quad x_{NO_xf}\right]^T \tag{6.24}$$

where now $x_{NO_x}(k) \in \mathbb{R}$ is added to track the actual NO_x, the model input $u_k = x_{NO_x}(k - \tau_{NO_x}/)$ is conveniently delayed and $z_k = z_{NO_x}(k)$. Matrices $\mathbf{F}$, B and H are modified accordingly

$$\mathbf{F} = \begin{bmatrix} 1 & 0 & 0 \\ 1 & 0 & 0 \\ 0 & 1 - a_{NO_x} & a_{NO_x} \end{bmatrix} \quad B = \begin{bmatrix} 0 \\ 1 \\ 1 - a_{NO_x} \end{bmatrix} \quad H = \begin{bmatrix} 0 \; 0 \; 1 \end{bmatrix} \tag{6.25}$$

Due to the sensor delay, the bias $\hat{\theta}_{NO_x}(k - \tau_{NO_x}/)$ at the instant k is observed with a certain delay τ_{NO_x}, and that means that at a given time, the actual drift will be calculated with the delay τ_{NO_x}

$$\hat{x}_{NO_x}(k) = x_{NO_x}(k) + \hat{\theta}_{NO_x}(k - \tau_{NO_x}/T_s) \tag{6.26}$$

However, the actual NO_x estimation should be

$$\hat{x}_{NO_x}(k) = x_{NO_x}(k) + \hat{\theta}_{NO_x}(k) \tag{6.27}$$

which corresponds to a non-causal operation.

Using the delayed $\hat{\theta}_{NO_x}(k - \tau_{NO_x}/)$ could lead to important errors when sharp load transients. Alternatively, an adaptive map $\mathbf{\Theta}_{NO_x} \in \mathbb{R}^2 : [n, m_f] \rightarrow \theta_{NO_x}$ could be introduced for modelling the drift as a function of the operating point conditions, as was made for λ^{-1}. This option is well described in Chap. 5: a map $\mathbf{\Theta}_{NO_x}$ is adapted by means of a steady-state version of a KF (similar to a RLS filtering) and is interpolated in a 2D basis for giving an initial estimate in the prediction of the state, in such way that the KF is modified as follows

$$\hat{x}_k = \mathbf{F}\mathbf{\Theta}_{NO_x,\mathbf{k}}(n(k), m_f(k)) + Bu_k + K_\infty\left(z_{NO_x}(k) - H\left(\mathbf{F}\hat{x}_{k-1} + Bu_k\right)\right) \tag{6.28}$$

and the table $\boldsymbol{\Theta}_{NO_x}$ is independently updated from the KF in order to store θ_{NO_x}

$$\boldsymbol{\Theta}_{\mathbf{NO_x,k}} = f\left(\boldsymbol{\Theta}_{\mathbf{NO_x,k-1}}, \hat{x}_k, z_{NO_x}(k)\right) \tag{6.29}$$

The map itself is interpolated for inferring θ_{NO_x} avoiding the causal problem and huge integrations when the drift varies with the operating point conditions and time. The final estimation is then calculated as follows

$$\hat{x}_{NO_x}(k) = x_{NO_x}(k) + \boldsymbol{\Theta}_{\mathbf{NO_x,k}}(n(k), m_f(k)). \tag{6.30}$$

6.3.3.2 Actual NO_x by Updating Look-up Tables

The state-space model for learning maps is useful not only for calibrating the map but also for the RT observation of the NO_x signal. Therefore, the matrices and the state vector in (5.33) are also modified and augmented for introducing x_{NO_x}

$$X^w = \left[vec(\boldsymbol{\Theta}_{NO_x}) \;\; x_{NO_x} \;\; x_{NO_xf}\right]^T \tag{6.31}$$

with the local observable matrix $\mathbf{F_k^o}$

$$\mathbf{F_k^o} = \left[\begin{array}{c|cc} & 0 & 0 \\ \mathbf{I_{4\times4}} & 0 & 0 \\ & 0 & 0 \\ & 0 & 0 \\ \hline C_{mod}\, q_k & 0 & 0 \\ 0 & (1-a_{NO_x}) & a_{NO_x} \end{array}\right] \tag{6.32}$$

For the case of the algorithm B for updating only look-up tables, $C_{mod} = 1$ ((model (4.8)). For the case of an algorithm like C and for the NO_x model,

$$C_{mod} = e^{-k_{NO_x}\cdot(EGR\lambda^{-1}-EGR_0\lambda_0^{-1})} \cdot C_{NO_x} \cdot C_{dyn} \tag{6.33}$$

with the parameters calibrated offline. Finally, H^w is now constant for both algorithms B and C

$$H^w = [0 \;\; 0 \cdots 1] \tag{6.34}$$

The use of the table allows to solve the non-causalities related with the RT estimation in a smart way: note that the observer is built on the basis of delayed inputs.

The use of a drift correction model or adaptive look-up tables NO_x leads to similar results, but the look-up tables structure is beneficial as far as can be used also for updating the model. In order to decouple the filter tuning for updating the look-up table and the output drift correction, the two methods could be combined by observing both the table parameters, the bias and the filtered output, designing a state-space vector

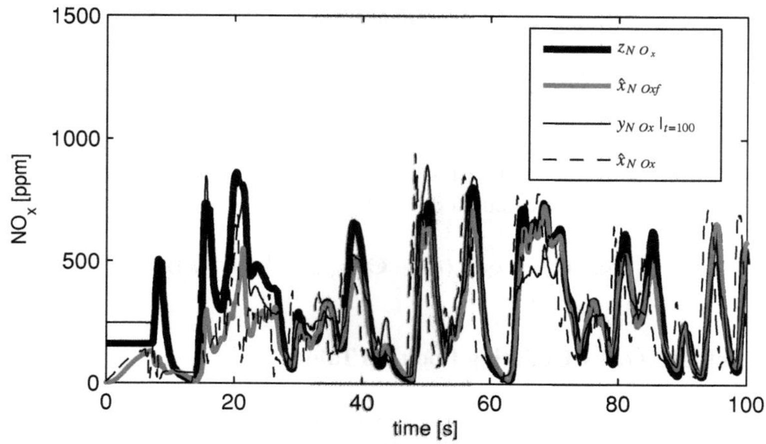

Fig. 6.19 NO_x estimation by three ways: z_{NO_x} sensor measurement, online observation of the sensor signal by the SKF method $\hat{x}_{NO_xf}$, offline prediction y_{NO_x} by using the map $\boldsymbol{\Theta}_{NO_x,100}$ and online actual estimation $\hat{x}_{NO_x}$

$$X^w = \left[vec(\boldsymbol{\Theta}_{NO_x}) \quad \boldsymbol{\theta}_{NO_x} \quad \mathbf{x}_{NO_x} \quad \mathbf{x}_{NO_x\mathbf{f}}\right]^T . \tag{6.35}$$

6.3.3.3 Results

The model with the state-space vector (6.31) is observed in order to show an application example of the fast estimation of the actual NO_x, taking into account that this algorithm can be adapted to other applications. Therefore, the look-up tables learning structure can be utilised for updating models and also estimating actual signals. Figure 6.19 shows results in the cycle SDMP B5, by fitting the NO_x model $NO_{x,0}$ map with initial null values and keeping the calibration set (6.14). $\hat{x}_{NO_xf}$ represents the observation of z_{NO_x} by using the adaptive map at every iteration. The adaptive map provides a perfect fitting in about 30 s (it might be accelerated varying the filter calibration but it is not advisable if the sensor model is not perfectly identified). Alternatively, $y_{NO_x}|_{t=100}$ shows the offline NO_x sensor prediction using the model $\boldsymbol{\Theta}_{NO_x,100}$ with acceptable results.

The table itself might predict the actual NO_x emissions if interpolating the table directly applying no delay nor filtering. Furthermore, the actual NO_x is directly observed with the state-space system for learning having a compact programming and providing an adaptive estimation. For that, x_{NO_x} is included in the state vector. This does not compromise the assumptions made for the SKF. Coming back to Fig. 6.19, $\hat{x}_{NO_x}$ is the online observation of the actual NO_x. This signal is advised for using it as a valid reference of the actual NO_x.

6.4 Conclusions

This section has presented different applications for the methods developed in Chap. 5 applied to the fast estimation of λ^{-1} and NO_x.

For the case of λ^{-1}, a complete procedure for the drift correction of a λ^{-1} model is presented by programming an augmented model. The robustness against signal uncertainties, coming from sensors or models, is ensured by introducing the so-called freezing rules that deactivate the updating when some conditions are fulfilled. A map for modelling the bias is proposed in order to solve the drift problem and to overcome the temporal and operating point variability of the bias.

The second part of the chapter emphasises the fast estimation of NO_x by following different methods.

First, the only use of look-up tables and a model sensor for estimating NO_x is presented. This section completes the experimental validation of the different methods for updating look-up tables, presented in Chap. 5. The results with all the methods are acceptable, but due to the robust properties of the SKF, this method is selected for the rest of the work.

Second, different observers are proposed in order to the online adaptation of the NO_x model presented in Chap. 4: the λ^{-1} observation, the online updating of the nominal $NO_{x,0}$ table and the model bias correction. The results show that the model bias can be tracked by the online updating of the model parameters and inputs.

Finally, the estimation of the actual NO_x by avoiding the delay and sensor filtering is proposed by augmenting the state-space models in order to observe the actual signal. Concretely, two alternatives are proposed: a drift correction model and a look-up tables learning model. The results are presented and validated by utilising the second option.

References

Arrègle J, López JJ, Guardiola C, Monin C (2008) Sensitivity study of a NO_x estimation model for on-board applications. SAE Tecnhical Report 2008–01-0640

Arsie I, Di Leo R, Pianese C, De Cesare M (2012) Combustion noise and pollutants prediction for injection pattern and EGR tuning in an automotive common-rail diesel engine. ECOSM 2012 workshop on engine and powertrain control, simulation and modeling

Desantes JM, Luján JM, Guardiola C, Blanco-Rodriguez D (2011) Development of NO_x east estimate using NO_x sensors. In EAEC 2011 congress, Valencia

Guardiola C, Pla B, Blanco-Rodriguez D, Cabrera P (2013a) A learning algorithm concept for updating look-up tables for automotive applications. Math Comput Model 57(7–8):1979–1989

Guardiola C, Pla B, Blanco-Rodriguez D, Eriksson L (2013b) A computationally efficient Kalman filter based estimator for updating look-up tables applied to NO_x estimation in diesel engines. Control Eng Practice 21(11):1455–1468

Guardiola C, Pla B, Blanco-Rodriguez D, Mazer A, Hayat O (2013c) A bias correction method for fast fuel-to-air ratio estimation in diesel engines. Proc Inst Mech Eng J Automob Eng 227(8):1099–1111

Höckerdal E, Frisk E, Eriksson L (2011) EKF-based adaptation of look-up tables with an air mass-flow sensor application. Control Eng Practice 19:442–453

Payri F, Guardiola C, Blanco-Rodriguez D, Mazer A, Cornette A (2012) Methodology for design and calibration of a drift compensation method for fuel-to-air ratio estimation. SAE Technical Paper 2012–01-0717

Polóni T, Rohal'-Ilkiv B, Alberer D, del Re L, Johansen TA (2012) Comparison of sensor configurations for mass flow estimation of turbocharged diesel engines, vol 418. Lecture notes in control and information sciences

Schilling A (2008) Model-based detection and isolation of faults in the air and fuel paths of common-rail DI diesel engines equipped with a lambda and a nitrogen oxides sensor. PhD thesis, ETH-Zürich

Schilling A, Amstutz A, Onder CH, Guzzella L (2006) A real-time model for the prediction of the NO_x emissions in DI diesel engines. In: Proceedings of the 2006 IEEE international conference on control applications, Munich, Germany

Tschanz F, Amstutz A, Onder CH, Guzzella L (2012) Feedback control of particulate matter and nitrogen oxide emissions in diesel engines. Control Eng Practice 21:1809

Part IV
Conclusions and Future Works

Chapter 7
Conclusions and Future Works

7.1 Main Contributions and Conclusions

The present dissertation covers the topic of the online dynamic estimation of λ^{-1} and NO_x in diesel engines. For that purpose, different sources of information are utilised:

- physical sensors for measuring exhaust gas concentrations (and intake oxygen when necessary), and
- control oriented models using ECU signals as fast inputs for estimating gas concentrations.

By combining the sensor and models, data fusion strategies are designed by means of observers based on the Kalman filter (KF) for the adaptive estimation of engine variables.

Specific tools, covering the online characterisation of sensors, the design of fast models and adaptive observers are developed in this work. Even though the methods of this work are applied to the estimation of λ^{-1} and NO_x, these can be applied to other engine variables, such as other gas concentrations or the volumetric efficiency, if a set-up with sensors and models is available. In the following, the main contributions and conclusions are presented and organised according to the thematic.

7.1.1 Online Characterisation of Gas Concentration Sensors

The measurement of engine-out gas concentration variables of a diesel engine is subjected to gas dilution, transport delays and is affected by the sensor hardware. In this work, a NO_x–ZrO_2 sensor is used for measuring λ^{-1} and NO_x at the exhaust line of a diesel engine (while a gas analyser and a UEGO sensor are used for calibration and comparison when necessary). NO_x sensors are useful for control and diagnosis of AT systems, but the specific characteristics of the dynamic response must be

D. Blanco-Rodriguez, *Modelling and Observation of Exhaust Gas Concentrations for Diesel Engine Control*, Springer Theses, DOI: 10.1007/978-3-319-06737-7_7,

assessed (Galindo et al. 2011; Guardiola et al. 2013b, c). The methods for the dynamic characterisation used in this work are based on the application of step-like transitions of the objective variables. For that purpose, ECU actuators are used for generating these profiles, allowing the on-board implementation of these methods.

7.1.1.1 Static Characterisation of λ^{-1} and NO_x

Both λ^{-1} and NO_x outputs of the ZrO_2 sensor are compared with those of a gas analyser, used as standard reference for the signals. The static accuracy, after measuring a representative set of the engine operating conditions, is enough for engine control purposes. A linear correlation is calculated with the static results.

7.1.1.2 Dynamic Characterisation of the NO_x Output

A simple on-engine experiment is proposed for the NO_x output characterisation. The method uses the engine as a gas generator and consists of applying steps in the start of injection (SOI) producing sharp variations (ΔNO_x) in the NO_x concentration. The use of SOI is justified by the high sensitivity of NO_x to SOI variations; the fact that SOI variations produce low variations in temperatures and pressures; and that SOI application is nearly instantaneous and no more than a cycle-to-cycle response time is expected. Furthermore, the SOI command can easily be varied in the diesel engine.

The dynamic response of the sensor is identified with a delayed first order filter. The main parameters, namely response time and delay, are identified by utilising least squares (LS) tools. The response time of the sensor is highly constant whatever the conditions, even though some variations appear at low NO_x and/or low ΔNO_x conditions. The identified delay has a bigger dispersion, especially when also approaching to idling. However the supposition of a constant delay for the designed algorithms gives satisfactory results.

7.1.1.3 Dynamic Characterisation of the λ^{-1} Output

A similar procedure to that of the NO_x output is made by performing injection step transitions provoking fast responses of λ, and thus λ^{-1}. LS methods are also used for identifying the delay and response time of a first order filter that models the sensor response.

The conclusions reached for the NO_x output are also applicable to the λ^{-1} output. The difference here is that the responses are a little bit faster, both in terms of delay and response time, because the output generated by an oxygen ion pump is located in a first cavity closer to the engine runner; while the NO_x output is proportional to a second oxygen ion pump located in a second chamber, thus increasing the delay and filtering.

In this work, a constant response time and delay have been utilised for modelling both NO_x and λ^{-1} outputs from the NO_x sensor. The results with the applied algorithms justify the selection of these simplistic sensor models.

7.1.2 Control Oriented Models for λ^{-1} and NO_x

The present dissertation designs virtual sensors for the on-board prediction of λ^{-1} and NO_x by means of fast models which take profit of signals coming from the ECU.

7.1.2.1 Fast Modelling of λ^{-1}

The relative fuel-to-air ratio (λ^{-1}) is estimated on the basis of the ECU signals for the injected fuel mass flow ($\dot{m}_f$) and the air mass flow ($\dot{m}_a$) (Payri et al. 2012a; Payri et al. 2012b; Guardiola et al. 2013c). Both signals are fast, which in principle guarantees the dynamic response of the model. For simplicity, this model neglects the mass accumulation effects that affect the intake and exhaust manifolds. Such effects could be considered for correcting the value $\dot{m}_a$ during the engine transients, e.g. when the boost pressure (p_{boost}) is increased.

The model presents significant errors in the steady-state determination of λ^{-1}. This is due to the deviations that exhibit the hot wire sensor used in the series engine for measuring $\dot{m}_a$, and because the fuel mass is not measured, but determined through the rail pressure and the injection duration.

The bias on $\dot{m}_a$ is expected to vary with the speed (because of the fluctuation patterns in the manifolds) and load (different set-points for p_{boost} and $\dot{m}_a$). Possible leaks in the manifold and blow-by also depend on the engine load, thus affecting the total signal error. On the other hand, the error in $\dot{m}_f$ estimation is affected by the injection settings (and hence load and engine speed).

When comparing the dynamic performance of the model with that of the sensor, the model exhibits almost instantaneous response: $\dot{m}_f$ estimate, although biased, is instantaneous; and $\dot{m}_a$ characteristic response time is in the order of milliseconds.

7.1.2.2 Fast Modelling of NO_x

An ECU oriented model for estimating NO_x is designed based on an exponential variation around the nominal behaviour as a function of the available intake oxygen and a set of corrections for catching possible deviations from the nominal. First principle equations function of different ECU signals, such as n, m_f, $\dot{m}_a$, p_{boost} and the engine coolant temperature (T_{cool}), are combined with different maps used for modelling the steady-state behaviour of the engine (Guardiola et al. 2013b).

The available intake oxygen is approximated by means of the product of the EGR rate and λ^{-1}. A calculation based on the volumetric efficiency map, and the $\dot{m}_a$ and p_{boost} signals, is utilised for estimating the EGR rate, after computing the intake temperature by a mixing model. The exhaust temperature (T_{exh}) is modelled by tables and used for inferring the EGR manifold temperature (T_{egr}). A set of corrections (C_{NO_x}) for coping with different deviations in the intake mass flow ($\dot{m}_{int}$), the intercooler temperature (T_{wic}), ambient humidity (H) and the intake temperature (T_{int}) is proposed. The effect of the thermal transients is considered by means of a factor C_{dyn} that considers the temperature in the cylinder walls by means of a dynamic equation depending on the thermal loading, characterised by the product $\lambda^{-1}n$.

The model is tuned by means of steady-state tests for filling the maps and parameters, while it is also validated by means of dynamic engine cycles. The identified sensor model for the NO_x output of the ZrO_2 sensor is used for comparing the model with the sensor output.

The results are satisfactory with both static and dynamic cycles, showing the capabilities of the model. The dynamic assumptions are sufficient for getting acceptable dynamic responses. The model tuning (mainly maps) is dependent on the calibration data set, and the ageing and changing conditions affect the model output, producing a biased response.

7.1.3 Observers for the Fast Estimation of Engine Variables

Two different observers are proposed for being used as adaptive estimators of engine variables:

- augmented models for drift correction, and
- on-board methods for updating look-up tables.

It is worth noting that these methods, presented in Chap. 5, are suited for a general problem, and can be applied for other engine variables if required inputs are available.

7.1.3.1 Augmented Models for Drift Correction

The data fusion is done through a steady-state KF for observing the bias between a fast model affected by drift and a slow but accurate sensor (Payri et al. 2012a; Payri et al. 2012b; Guardiola et al. 2013c). The filter works as a pass filter that keeps the high frequency components of the model and the low frequencies of the sensor. The model states are augmented for tracking the model bias. The steady-state analytical solution of the Kalman gain is solved and computed beforehand as function of the response time of the sensor and a noise quotient between the process and sensor noises. Guidelines for the observer tuning are proposed by a sensitivity analysis of the estimations covering possible errors or variations in the sensor modelling.

7.1.3.2 Methods for Updating Look-up Tables

The problem of updating look-up tables by methods based on the extended Kalman filter (EKF) is treated (Guardiola et al. 2013a, b). The table parameters are treated as states that can be observed if an output measurement is given. The table output is computed with a first order linear interpolation, which defines the output matrix of the learning state-space system. In a first approach, the problem has been focused on the use of a static table whose output is directly the aimed variable, i.e. the sensor dynamics are not considered in a first approach.

The EKF, with respect to other recursive methods for identification such as the recursive least squares (RLS), presents the advantage of tracking the estimation error by means of a covariance matrix. Nevertheless, when coping with large systems, the computational burden is high, in terms of memory, fact that makes difficult its implementation for on-board strategies.

Beyond the EKF and based on it, two methods are developed:

- A steady-state KF (SSKF) calculated by means of building independent state-space models for every relative position of the matrix, which corresponds to a given operating point, e.g. scheduled by n and m_f in a general case. Therefore, if the inputs are considered stationary, the KF is steady-state under these hypotheses and a constant Kalman gain can be derived (the filter is linear time invariant). If covering the whole grid, the Kalman gain can be mapped for the operating range, depending on the inputs and noise trade-off.

  The analytical solution of the steady-state filters is demonstrated and a proportional correction method is developed, resulting in a fast filter, whose gains can also be mapped beforehand if the noise variances are constants. Even though the method does not track the covariance error, the magnitude of the corrections could be modified online just varying the noise trade-off and recalculating the Kalman gain (by the analytical functions of by interpolating the map). The method behaves well for data when variations are rapid, closer to random data. However, it presents some convergence problems when the system knowledge is poor and the excitation level is not sufficient.
- A simplified KF (SKF) that lights the computation of the covariance matrix of the KF. This method exploits the fact that a N-D look-up table has a structure where at every iteration a maximum number of 2^N elements are active, allowing to skip the updating of the elements out of this area. Nevertheless, a KF tracks the ageing of all elements, indeed the inactive ones, by means of storing large covariance matrices and compromising the memory and processing. The SKF method neglects the non-diagonal relationships of the covariance matrix (crossed covariances between table elements), but keeps the updating of the diagonals (variances).

  The SKF method requires similar resources to those of the SSKF, but the performance is more similar to the one of the KF. The solution of the SKF is robust and the method behaves well under different data settings. The SKF is well-suited for a robust on-board implementation of look-up table estimators.

The static look-up tables can be linked with first order filters for representing dynamic equations, related with the sensor measurement or physical processes. For including the filters in the learning structure, an augmentation of the state-space model is proposed. After validating the methods by learning an adaptive table that models the NO_x emissions, the SKF method is advised instead of the KF (heavier) or the SSKF (less robust), due to its performance is similar to that of the KF but with a comparable computational cost to that of the SSKF.

7.1.4 *Application of Adaptive Estimators to Infer λ^{-1} and NO_x in Diesel Engines*

Considering the methods developed in Chap. 5, the following three algorithms present useful data fusion strategies:

- algorithm A for tracking bias by an augmented model and a steady-state KF estimator,
- algorithm B for directly updating a static look-up table, and
- algorithm C for including the algorithm B structure inside a model for updating a table or parameter of a given model.

The results with the proposed structures are shown by applying to the engine-out λ^{-1} and NO_x signals in a diesel engine. The uncertainties on the sensors and the models are considered as well.

7.1.4.1 Fast Observation of λ^{-1}

An observer is initially proposed for tracking the bias on the λ^{-1} signal computed on the basis of the fuel-to-air ratio (Guardiola et al. 2013c). A set of rules is established in order to stop the bias integration when certain conditions do not ensure the problem robustness, which are:

- uncertainties on the sensor model when sharp transients,
- saturation problems of the sensor, and
- unexpected behaviours due to inaccuracies in the model or involved sensor signals, i.e. m_f, $\dot{m}_a$, λ^{-1}.

The estimation is accurate, even though two problems are underlined. First, the causality of the solution since the sensor response is delayed and therefore the actual bias is not computed on real time; second, the bias varies with the operating point conditions which involve high variations of the state.

These problems are solved by introducing an adaptive look-up table for modelling the bias. The table elements and the bias can be included directly in the observer structure. However, the table is kept as an independent block of the drift correction, being able to program independent filters for the bias and the table. Both the SSKF and the SKF methods are suitable for updating the table, although the results are shown by applying the former. The table output gives the prediction for the bias, while a steady-state KF is applied to update this bias and at the same time for giving the reference for also updating the table, and so on.

7.1.4.2 Fast Observation of NO_x

The three different algorithms (A, B and C), and some modifications based on them, are applied to the estimation of NO_x emission in diesel engines.

Adaptive look-up tables for modelling NO_x
NO_x emissions are modelled first by introducing a static 2D look-up table, as function of n and m_f in order to catch the nominal NO_x, and a first order filter for representing the sensor dynamics (Guardiola et al. 2013b).

The three learning methods (EKF, SKF and SSKF) are implemented and the results compared in order to fill the table parameters without any previous knowledge. The results are satisfactory for the three methods, but due to the special properties of the SKF, this method is advisable.

These tables can model with acceptable accuracy NO_x emissions when the cycles are slow enough, allowing to simplify the closed loop engine response to a function that depends uniquely on n and m_f. Under this condition, the learning on a given cycle, i.e. NEDC, can be used for representing NO_x emissions in other similar cycles such as the CADC, provided that the table elements have already been updated.

Online updating of the NO_x *model*
Different modifications are introduced to improve the online and offline capabilities of the NO_x model: the observation of λ^{-1} when a sensor signal is available, the adaptation of the main table of the NO_x model ($NO_{x,0}$), and the possible state augmentation by tracking the model bias (discussed in the next paragraph). These modifications are proposed for correcting the ageing while rejecting spurious signals.

The benefits of the λ^{-1} observation are clear from the λ^{-1} estimation, avoiding that incorrect model estimations affect the NO_x model output. A drift correction algorithm is proposed when the NO_x sensor (or a UEGO sensor) is available.

The use of a learning model for updating the $NO_{x,0}$ table in the NO_x model allows to avoid the ageing, minimising the online estimation error and the offline prediction error (model is adaptive). The SKF method is used for the application. The updating method is useful for both fitting a table when no initial knowledge exists or updating the model for cancelling the drift.

Online observation of the actual NO_x
The actual NO_x is inferred on the basis of the adaptive filtering models. Two alternatives are designed in this work.

First, a drift correction model is proposed for tracking the bias, and thus correcting the filtered NO_x model output when a NO_x sensor is available. This model is augmented with a state for tracking the actual NO_x (before filtering and delaying), resulting from adding the bias to the NO_x model output. The algorithm and causality problems are similar to those of the λ^{-1} estimation.

Second, the learning model for look-up tables can also be augmented with the actual NO_x state. In this case, the structure avoids the causality problem when the RT estimation is required. The direct interpolation of the table is the actual NO_x output, when the only delays are associated with the ECU signals, mainly n, m_f, $\dot{m}_a$ and p_{boost}.

A further augmentation of the model with a bias state makes possible to decouple the table updating with respect to the fast actual NO_x estimation, in such a way that the related noises for the bias and table elements are tuned separately:

- Table parameters may be updated slowly by selecting a low noise related to those parameters,
- while the bias can be integrated faster if the sensor model is reliable and hence selecting a bigger noise for this state.

Other different trade-off might be selected in order to speed up the table convergence, but this structure allows to decouple in some way the table updating and the bias tracking. Of course that the lower the model error (table output) the lower the bias.

7.2 Future Works

Even though the results presented in this work prove the validity of the proposed methods, the on-board implementation of them still deserves a deeper study. Particularly, the works to be performed as a continuation of this dissertation can be divided into those related with possible improvements of the sensor models, the design of an adaptive MVEM of the engine air path, the design of more complex emission models (and their possible adaptation), the validation of the methods and the different applications of the algorithms.

7.2.1 Sensor Knowledge

A deeper study on the NO_x sensor response is mandatory in order to identify a physical model that can predict both the delay and response time with higher accuracy. The sensor delay and response times are variable, especially at low loads and low NO_x emissions [partly due to the loss of accuracy of the sensor for low NO_x concentrations (Groß et al. 2012)]. Different phenomena and effects deserve an attention: the protective layer distortion, the internal diffusion speed between chambers and the effect of flux velocities, pressures, temperatures and concentrations over the final sensor output. Even though the pressure effect over the NO_x output from the NO_x sensor has been identified and it is not high for the pressures that can appear at the diesel engine exhaust, a deeper study should be conducted in order to characterise an appropriate correction. Furthermore, the influences of the pressure and other operation variables over the sensor should also be analysed.

To sum up, the use of a first order filter model with constant response times and delay could be improved by considering possible variations in these parameters as well as other influences that can be modelled for instance with transfer functions, or alternatively by considering a more representative non-linear model. In order to fulfil this task, the use of fast response measurements is required for calibration and validation.

7.2.2 Design of an Adaptive MVEM for the Air Path

The air path model used for computing the EGR mass flow entering the engine is based on the volumetric efficiency and the measurements of the air mass flow and the boost pressure. The influence of the storage dynamics at the intake and the exhaust junction are not considered nor the dynamic effects over the intake and exhaust temperatures. Furthermore, the exhaust temperature is also modelled by a map with a set of corrections. The model could also consider the turbocharger (TC), the air filter, the actuators or the intercooler as well as the after-treatment (AT) systems.

The design of a MVEM considering these effects is not a challenge (this topic is well covered in the bibliography and the state of the art) but the adaptation of a MVEM is an interesting problem. Different strategies, similar to those presented in this work, can be implemented in order to estimate model states and inputs [see e.g. (Polóni et al. 2012)].

With regards to the observation of the model inputs and states, drift correction models can be set for tracking the bias between the model states and sensor measurements. Good examples are the drifts on $\dot{m}_a$ (by considering a model for the TC) and the injector (by using λ^{-1} and the corrected $\dot{m}_a$).

With respect to the observation of model parameters as states, the fusion strategies can be utilised for updating the model while also for fitting the ECU parameters. Hence the drift correction and the updating algorithms are useful tools. However, there exists a problem with respect to the observability of the states and the filter tuning, i.e. it is difficult to weight the errors between all the parameters and the robustness must be assessed. The problem could be split up by designing local observers for the different sub-models, e.g. observing compressor maps by comparing with the MAF sensor output. Anyway, the closed-loop interaction should be assessed and a global cost function optimised.

This problem deserves a specific study and the use of synthetic signals should be a first step for evaluating the possibilities of the observers for solving a global optimisation problem. When involving look-up tables (frequent on MVEMs and ECUs), methods like the SKF permit the online adaptation with an affordable computational effort.

If the updating problem is designed globally and considering the closed loop interaction on real time, the discussion made for the SKF in order to simplify the KF when the system is large and only a few parameters are observable is also valid. Here, the use of parameters for defining how observable elements are at every iteration (e.g. grammians) allows to simplify the observation problem.

7.2.3 Emission Models

The designed NO_x and λ^{-1} models are based on static maps and equations with a set of corrections and ECU signals as inputs, plus filters and delays for modelling physical and sensor dynamics. If other sensors are available, such as a cylinder pressure sensor, the prediction of the exhaust conditions can be benefited from a more accurate combustion model. The model states could also be augmented by including the soot prediction and considering further states from a MVEM of the air path.

In addition, an interesting improvement is modelling the AT devices. For instance, the monitoring of soot upstream and downstream of the DPF and the tight control of the exhaust λ^{-1} is required for an optimal control of the DPF. For a correct estimation, the combination of sensors and models with similar methods to those developed in this dissertation is advisable. The same example can be applied to SCR.

The physical interactions should also be included in control-oriented models of the AT line (Guzzella and Sciarretta 2007). This is of particular importance for tracking the NO_x emissions at different points, i.e. although the total NO_x could be highly constant upstream of the deNO_x system (SCR or LNT), the concentration of NO and NO_2 species vary due to the oxidation in the DOC. And of course, the total NO_x will be reduced after deNO_x systems. The crossed effects over other concentrations (e.g. soot) and over the sensor signals, as well as other factors such as the ammonia cross sensitivity should be taken into account. For such cases, the data fusion of sensors and models by observers is also interesting, as shown in Hsieh and Wang (2011).

7.2.4 Validation of the Models and Adaptive Methods

The assessment of the estimations and sensors should be validated with fast response measurement systems that can offer a dynamic and accurate reference of the considered variables. They are necessary for improving the sensor model identification and the validation of the methods.

Fast gas analysers based on CLD for measuring NO_x present response times around 2 ms, those based on NDIR for measuring CO/CO_2 (thus oxygen or λ^{-1}) reach 8 ms, while for soot these can be lower than 100 ms. These devices are suitable in order to identify the sensor dynamics and validate the results.

7.2.5 Applications

Once these methods and models are validated, the RT estimation of engine states (NO_x and λ^{-1} in the present work) could be used for the following functions:

- Control of the air path, fuel path and AT devices with feedback on exhaust emissions and other engine states,
- OBD of the AT systems,
- serving as inputs for other models or functions, as for instance, soot models (which depends on the exhaust λ^{-1}), and
- tuning of controller parameters by model based optimisation (Arsie et al. 2011).

To sum up, if a proper sensor set is available, engine variables and parameters can be observed. In addition to λ^{-1} and NO_x, interesting examples could be the volumetric efficiency, the intake oxygen concentration, the mass flows, temperatures and pressures in the manifolds, and other relevant exhaust concentrations. Due to the EURO 6 and future legislations, this topic is of high interest, and recent advancements in sensor technology, as the case of soot sensors, may be benefited by the presented methods. Anyway, the final quality on the estimation also falls on the accuracy of the models used for the data fusion and the knowledge of the sensor behaviour.

References

Arsie I, Criscuolo I, Pianese C, De Cesare M (2011) Tuning of the engine control variables of an automotive turbocharged diesel engine via model based optimization. SAE Technical Paper 2011-24-0146

Galindo J, Serrano JR, Guardiola C, Blanco-Rodriguez D, Cuadrado IG (2011) An on-engine method for dynamic characterisation of NO_x concentration sensors. Exp Thermal Fluid Sci 35(3):470–476

Groß A, Beulertz G, Marr I, Kubinski DJ, Visser JH, Moos R (2012) Dual mode NO_x sensor: measuring both the accumulated amount and instantaneous level at low concentrations. Sensors (Basel) 12(3):2831–2850

Guardiola C, Pla B, Blanco-Rodriguez D, Cabrera P (2013a) A learning algorithm concept for updating look-up tables for automotive applications. Mathematical Comput Modelling 57(7–8):1979–1989

Guardiola C, Pla B, Blanco-Rodriguez D, Eriksson L (2013b) A computationally efficient Kalman filter based estimator for updating look-up tables applied to NO_x estimation in diesel engines. Control Eng Pract 21(11):1455–1468

Guardiola C, Pla B, Blanco-Rodriguez D, Mazer A, Hayat O (2013c) A bias correction method for fast fuel-to-air ratio estimation in diesel engines. Proc Inst Mech Eng Part D: J Automobile Eng 227(8):1099–1111

Guzzella L, Sciarretta A (2007) Vehicle propulsion systems. Introduction to modeling and optimization, 2nd edn. Springer, Heidelberg (ISBN 978-3-540-74691-1)

Hsieh M-F, Wang J (2011) Design and experimental validation of an extended Kalman filter-based NO_xconcentration estimator in selective catalytic reduction system applications. Control Eng Pract19(4):346–353

Payri F, Guardiola C, Blanco-Rodriguez D, Mazer A, Cornette A (2012a) Methodology for design and calibration of a drift compensation method for fuel-to-air ratio estimation. SAE Technical Paper 2012-01-0717

Payri F, Guardiola C, Pla B, Blanco-Rodriguez D, Mazer A, Cornette A (2012b) Fast estimation of fuel-to-air ratio through the combination of a fast model and an exhaust lambda sensor. In: SIA 2012 International Conference

Polóni T, Rohal'-Ilkiv B, Alberer D, del Re L, Johansen TA (2012) Comparison of sensor configurations for mass flow estimation of turbocharged diesel engines, vol 418. In: Proceedings of Lecture Notes in Control and Information Sciences

Appendix A

A.1 Gas Concentration Sensor Specifications

Table A.1 presents the relevant characteristics of the gas concentration sensors certified by the OEMs.

A.2 Principles of the NO_x Measurement

The term NO_x emissions includes the di-nitrogen oxide N_2O, nitrous oxide NO and the nitrogen oxide NO_2 (Turner 2009); the most habitual at the engine exhaust in diesel engines is NO, with percentages about 70–90 % (Hernández 2004; Schilling 2008), while the remainder is mainly NO_2.

The gas analyser is capable of measuring NO_x and separating between NO and NO_2 if necessary. For the NO_x case, the total composition is first oxidised to NO_2 and then converted by a catalytic reaction to NO, that afterwards reacts with ozone (O_3) for producing NO_2 and emitting photons. The produced light is detected with the CLD and is proportional to the NO concentration. For detecting NO (before oxidising), the thermal catalytic phase is avoided.

The measurement provided by the gas analyser for NO_x is made in a dry basis and for converting to a standard wet basis, the correction K_w is calculated as follows

$$K_w = (1 - F_{fh}\frac{\dot{m}_f}{\dot{m}_a}(1 + 0.001\ H_a)) - K_{w2} \tag{A.1a}$$

$$K_{w2} = \frac{1.608\ H_a}{1,000 + 1.608\ H_a} \tag{A.1b}$$

$$F_{fh} = \frac{1.969}{1 + \frac{\dot{m}_f}{\dot{m}_a}} \tag{A.1c}$$

where $\dot{m}_a$ and $\dot{m}_f$ are the air mass flow and the injected fuel mass flow [kg/h] respectively, while H_a is the absolute humidity.

D. Blanco-Rodriguez, *Modelling and Observation of Exhaust Gas Concentrations for Diesel Engine Control*, Springer Theses, DOI: 10.1007/978-3-319-06737-7,

Table A.1 Relevant characteristics of the gas concentration sensors used for this dissertation

	Horiba MEXA 7100DEGR	Bosch LSU 4.9	Uni-NO_x sensor
Output signals	CO, CO_2, NO_x, THC, O_2	λ	NO_x, linear λ, binary λ
Measuring range	NO_x: 0–5,000 ppm	(I_p) 0.65 to air	NO_x: 0–1,500 ppm
	O_2: 0–25 % vol.		lin λ: 0.75 to air
	CO_2: 0–12 % vol.		bin λ: >0.75 V ($\lambda = 0.9$)
			bin λ: <0.2 V ($\lambda = 1.1$)
Accuracy	NO_x: ±1.5 % FS	$\lambda = 1.016 \pm 0.007$	NO_x: 100 and 500 ppm ±10 %
	O_2: ±1.5 % FS	$\lambda = 0.8 \pm 0.01$	0 ppm ± 10 ppm
	CO_2: ±1 % FS	$\lambda = 1.7 \pm 0.05$	lin λ: $\lambda = 1$ ±6 (1,000/λ) fresh
			bin λ: 1.002 ± 0.008
Response times	NO_x: <900 ms	<100 ms*	NO_x: 750 ms
	O_2: <750 ms		λ: 550 ms
	CO_2: <800 ms		

*Data not specified by the OEM. FS is Flux Storage

Furthermore, the legislation fixes the factor K_{hd} for accounting with the ambient conditions

$$K_{hd} = \frac{1}{1 + A(H_a - 10.71) + B(T_{atm} - 298)} \quad \text{(A.2a)}$$

$$A = 0.309\frac{\dot{m}_f}{\dot{m}_a}(1 + 0.001H_a) - 0.0266 \quad \text{(A.2b)}$$

$$B = -0.209\frac{\dot{m}_f}{\dot{m}_a}(1 + 0.001H_a) + 0.00954 \quad \text{(A.2c)}$$

where T_{atm} [K] is the ambient temperature. Therefore, the NO_x output after corrections is computed

$$NO_x = K_w K_{hd} NO_x^{gas\ analyser} \quad \text{(A.3)}$$

A deeper explanation on the gas analyser procedures for measuring NO_x and separating components is in (Hernández 2004).

In the case of the NO_x sensor, the exhaust gas is diffused first to a first chamber, where the oxygen is measured, and then diffused again to the second chamber. In the first chamber, NO species are oxidised to NO_2, while in the second, total NO_x is reduced to oxygen and nitrogen ions. These oxygen ions are pumped out and are proportional to the total NO_x concentration. Note that the first oxidation step is required in order to get a correct measurement of total NO_x, i.e. NO_2 contains two oxygen atoms while NO contains only one and then the oxygen ions resulted from the dissociation could not be directly related with total NO_x. The static calibration was shown in Fig. 3.14, where the fitting between ZrO_2 sensor and gas analyser is quite acceptable.

A.3 Analytical Solutions to the Riccati Equations

The drift correction algorithm (Sect. 5.2.1) and the SSKF approach (Sect. 5.3.2) for updating look-up tables constitute steady-state versions of the KF. This special property drives to calculate a steady-state Kalman gain, allowing to precompute it and saving online resources. The Riccati iterative matrix system (2.13) becomes a matrix algebraic system and analytical expressions may be solved.

A.3.1 Drift Correction Model

The dynamic Riccati equation (2.13a, b, c) is steady-state as K and $\mathbf{P}$ converge after a certain number of iterations (Simon 2001), which depend on the initial values. Then, it is possible to derive an analytical solution as proved by Kalman and Bucy (1961). Some authors have given solutions to particular problems, such as (Hewer 1973; Nash 1967; Ekstrand 1983; Sudano 1995) or indeed computationally effective solutions as (Lainiotis et al. 1994). Here, an analytical solution to the Riccati matrix algebraic system (2.13a, b, c) considering the model (5.1a, b) is derived.

$\mathbf{P_{k-1}}$ matrix is

$$\mathbf{P}_{k-1} = \begin{bmatrix} p_1 & p_2 \\ p_3 & p_4 \end{bmatrix} \tag{A.4}$$

Operating (2.13a), $P_{k|k-1}$ results

$$\mathbf{P_{k|k-1}} = \begin{bmatrix} p_1 + \sigma_w^2 & (1-a)p_1 + ap_2 \\ (1-a)p_1 + ap_3 & q \end{bmatrix} \tag{A.5}$$

where

$$q = (1-a)((1-a)p_1 + ap_2) + a(1-a)p_3 + a^2 p_4 \tag{A.6}$$

Calculating K_k from (2.13b)

$$K_k = \begin{bmatrix} k_1 \\ k_2 \end{bmatrix} = \begin{bmatrix} \frac{(1-a)p_1 + ap_2}{q + \sigma_v^2} \\ \frac{q}{q + \sigma_v^2} \end{bmatrix} \tag{A.7}$$

From (2.13c)

$$\mathbf{P_k} = \begin{bmatrix} p_1 + \sigma_w^2 - \frac{((1-a)p_1+ap_2)((1-a)p_1+ap_3)}{q+\sigma_v^2} & ((1-a)p_1+ap_2)\frac{\sigma_v^2}{q+\sigma_v^2} \\ ((1-a)p_1 + ap_2)\frac{\sigma_v^2}{q+\sigma_v^2} & \frac{q\sigma_v^2}{q+\sigma_v^2} \end{bmatrix} \quad \text{(A.8)}$$

The Kalman filter is steady-state and observable and sequences $\{\mathbf{P_k}\}_{k=1}^{k=\infty}$ and $\{K_k\}_{k=1}^{k=\infty}$ are convergent. The system solution can be computed through solving the system that results from setting (A.4) and (A.8) as equal.

$$p_1 = p_1 + \sigma_w^2 - \frac{((1-a)p_1+ap_2)((1-a)p_1+ap_3)}{q+\sigma_v^2} \quad \text{(A.9a)}$$

$$p_2 = ((1-a)p_1+ap_2)\frac{\sigma_v^2}{q+\sigma_v^2} \quad \text{(A.9b)}$$

$$p_3 = ((1-a)p_1+ap_3)\frac{\sigma_v^2}{q+\sigma_v^2} \quad \text{(A.9c)}$$

$$p_4 = \frac{q\sigma_v^2}{q+\sigma_v^2} \quad \text{(A.9d)}$$

$$k_1 = \frac{(1-a)p_1+ap_2}{q+\sigma_v^2} \quad \text{(A.9e)}$$

$$k_2 = \frac{q}{q+\sigma_v^2} \quad \text{(A.9f)}$$

From (A.9b) and (A.9c), $p_2 = p_3$, which was already known as $\mathbf{P}$ is symmetric, and system (A.9a) really has 5 equations with 5 unknowns.

From (A.9a) and (A.9e)

$$\sigma_w^2 = k_1^2(q+\sigma_v^2) \quad \text{(A.10)}$$

and from (A.9f)

$$q+\sigma_v^2 = \frac{\sigma_v^2}{1-k_2} \quad \text{(A.11)}$$

and combining these two expressions,

$$k_1 = \frac{\sigma_w}{\sigma_v}\sqrt{(1-k_2)} \quad \text{(A.12)}$$

which relates k_1 to k_2.

q parameter (A.6) depends on $\mathbf{P}$ elements and must be re-arranged for the correct problem solution. Then, from the symmetry of $\mathbf{P}$ ($p_2 = p_3$), and considering (A.9b) and (A.9e)

$$p_2 = p_3 = \sigma_w\sigma_v\sqrt{1-k_2} \quad \text{(A.13)}$$

From (A.9d) and (A.9f)

$$p_4 = k_2\sigma_v^2 \tag{A.14}$$

and finally from (A.9b), (A.11) and (A.13)

$$(1-a)p_1 + ap_2 = \frac{p_2}{1-k_2} = \frac{\sigma_w\sigma_v\sqrt{1-k_2}}{1-k_2} \tag{A.15}$$

Substituting (A.13), (A.14) and (A.15) in (A.6)

$$q = (1-a)\,(1+a-ak_2)\sqrt{\frac{\sigma_w^2\sigma_v^2}{1-k_2}} + a^2k_2\sigma_v^2 \tag{A.16}$$

For convenience, (A.16) is set as function of the noise covariance trade-off σ_w^2/σ_v^2

$$q = \left((1-a)\,(1+a-ak_2)\sqrt{\tfrac{\sigma_w^2/\sigma_v^2}{1-k_2}} + a^2k_2\right)\sigma_v^2 = s\sigma_v^2 \tag{A.17}$$

defining parameter $s = f(\sigma_w^2/\sigma_v^2)$

$$s = (1-a)\sqrt{\frac{\sigma_w^2/\sigma_v^2}{1-k_2}}(1+a-ak_2) + a^2k_2 \tag{A.18}$$

Considering (A.9f), (A.12) with (A.17), K can be computed

$$k_1 = \sqrt{(1-k_2)\frac{\sigma_w^2}{\sigma_v^2}} \tag{A.19a}$$

$$k_2 = \frac{s}{s+1} \tag{A.19b}$$

The set (A.19) represents an analytical solution of the steady-state Kalman filter that may be solved iteratively. Moreover, (A.19) shows that K depends exclusively on the noise covariance trade-off σ_w^2/σ_v^2 given a certain system (5.1).

P is solved from (A.9a)–(A.9d) and (A.19)

$$p_1 = \sqrt{\frac{\sigma_w^2\sigma_v^2}{1-k_2}}\left(1+\frac{ak_2}{1-a}\right) \tag{A.20a}$$

$$p_2 = p_3 = \sqrt{\sigma_w^2(1-k_2)} \tag{A.20b}$$

$$p_4 = k_2\sigma_v^2 \tag{A.20c}$$

although is not really necessary for the algorithm implementation.

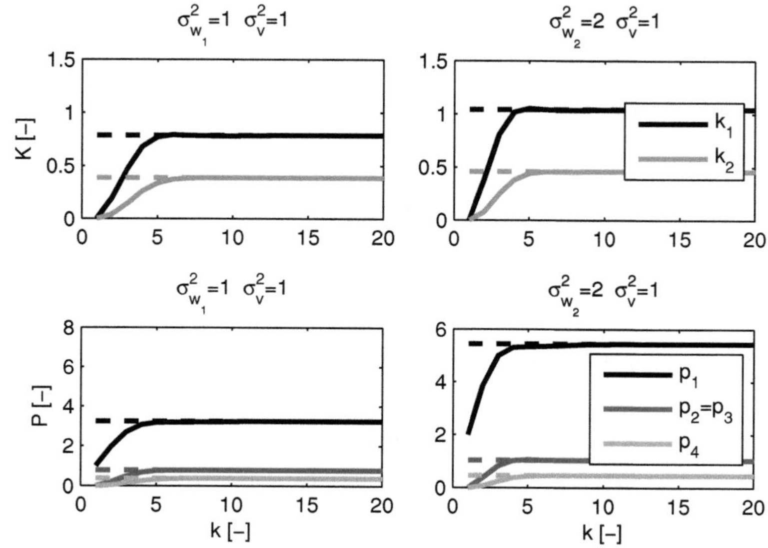

Fig. A.1 Kalman gain K and covariance matrix P calculation for two noise sets with $a = 0.8$. *Solid lines* corresponds to the standard Kalman formulation by the iterative Riccati equation (2.13a, b, c), while *dashed lines* correspond to the proposed analytical formulation ((A.19) for K and (A.20) for $\mathbf{P}$)

Numerical example

A numerical example using (A.19), (A.20) for calculating $\mathbf{P}$ and K of a steady-state Kalman filter is presented. The result must be equal to calculate the converged $\mathbf{P_k}$ and K_k from (2.13a, b, c) when the simulation time is high enough:

$$K = \lim_{k \to +\infty} K_k \tag{A.21a}$$

$$\mathbf{P} = \lim_{k \to +\infty} \mathbf{P_k} \tag{A.21b}$$

Parameter a is 0.8 and in all cases $\sigma_v^2 = 1$, while two different values are applied to σ_w^2: $\sigma_{w1}^2 = 1$ and $\sigma_{w2}^2 = 2$.

Figure A.1 shows the results applying the standard Kalman formulation and the ones calculated by the analytical formulation. Note that x axis shows the number of iterations (k), which only makes sense for the standard Kalman formulation as far as the proposed solution is calculated beforehand. Final values after convergence are exactly the same proving the validity of the proposed formulation.

A.3.2 The SSKF for Updating Look-Up Tables

Given the system (5.17) for updating look-up tables and a filter tuning σ_v^2/σ_w^2, the value of $\mathbf{P}$ does not converge under normal operation, nor does K_k. This is because H_k matrix is not constant and depends on the engine operating conditions. But if the operating point condition is constant, system runs in a constant point ($H_k = H$) with a certain σ_v^2/σ_w^2, then it is possible to derive a constant K^{SS}, although $\mathbf{P}$ does not really converge. This is observed from simulations and then it can be set out for a 2D grid

$$K^{SS}(\eta_1, \eta_2, \sigma_v^2/\sigma_w^2) = \lim_{k \to +\infty} K_k \tag{A.22}$$

Taking into account the symmetric property of the problem, the solution is set out for a 1D table $\mathbf{\Theta}$, whose elements are expanded in a state vector and whose 1D local observable has a state vector

$$x_k^o = [\Theta_i \quad \Theta_{i+1}]_k^T \quad \in \mathbb{R}^2$$

where i denotes the position of the first active element in $\mathbf{\Theta}$. The state-space equations of the local observable system are

$$x_k^o = \mathbf{I}_{2\times 2} x_{k-1}^o + w_k \tag{A.23a}$$

$$z_k = q_k^o x_k^o + v_k \tag{A.23b}$$

with

$$q_k^o = [1 - \eta \quad \eta] \tag{A.24}$$

and considering r as the vector that defines the 1D grid

$$\eta(k) = \frac{u_k - r_i}{r_{i+1} - r_i} \tag{A.25}$$

where $u_k \in \mathbb{R}$ is now the scheduling point considered as input to the table.

Noise covariances are

$$\mathbf{Q} = \begin{bmatrix} \sigma_w^2 & \\ & \sigma_w^2 \end{bmatrix}; \; R = \sigma_v^2 \tag{A.26}$$

By imposing the condition A.22, the value of Kalman gain k_i can be derived analytically with a certain effort

$$k_i(\eta, \sigma_v^2/\sigma_w^2) = \frac{0.5(1-\eta)(1+s)}{0.5(1+s)(1-2\eta+2\eta^2) + \sigma_v^2/\sigma_w^2} \tag{A.27}$$

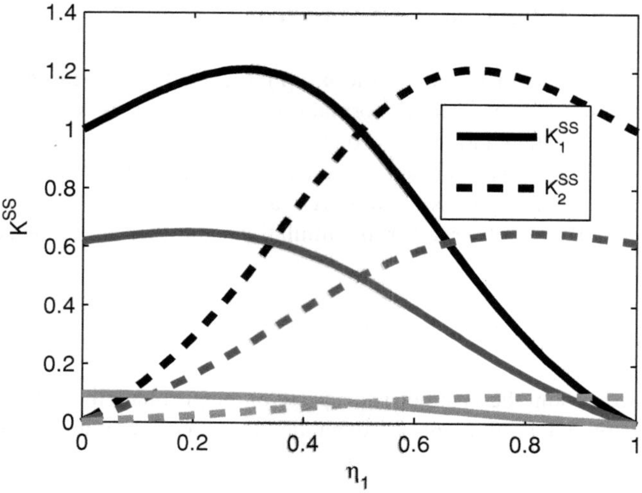

Fig. A.2 Membership functions for the Kalman gain correction for the 1D case for $\sigma_v^2/\sigma_w^2 = 0$ (*black*), $\sigma_v^2/\sigma_w^2 = 0.01$ (*medium grey*) and $\sigma_v^2/\sigma_w^2 = 0.1$ (*light grey*)

where

$$s = \sqrt{1 + \frac{4}{(1 - 2\eta + 2\eta^2)} \frac{\sigma_v^2}{\sigma_w^2}} \tag{A.28}$$

showing explicitly the dependance of k_i with the noise trade-off σ_v^2/σ_w^2 and the interpolation function defined by η. The result is used for calculating gain correction for the node i, when the observable region is between i and $i+1$ in the 1D case. By using symmetric properties

$$k_{i+1} = k_i(1 - \eta, \sigma_v^2/\sigma_w^2) \tag{A.29}$$

The function is plotted in Fig. A.2. For the 2D case, K^{SS} (see (A.22)) can be solved by considering at the same time the interpolation properties, and then the gain for a local observable system of a given 2D table with interpolation weightings η_1 and η_2 defined in (5.16) has the solution

$$K^{SS} = \begin{bmatrix} k_i(\eta_1, \sigma_v^2/\sigma_w^2) \cdot k_i(\eta_2, \sigma_v^2/\sigma_w^2) \\ k_i(\eta_1, \sigma_v^2/\sigma_w^2) \cdot k_i(1 - \eta_2, \sigma_v^2/\sigma_w^2) \\ k_i(1 - \eta_1, \sigma_v^2/\sigma_w^2) \cdot k_i(\eta_2, \sigma_v^2/\sigma_w^2) \\ k_i(1 - \eta_1, \sigma_v^2/\sigma_w^2) \cdot k_i(1 - \eta_2, \sigma_v^2/\sigma_w^2) \end{bmatrix} \tag{A.30}$$

The obtained solution can be easily generalised for the nD case by solving

$$\prod_{i=1}^{n} k_i(\xi_i, \sigma_v^2/\sigma_w^2) \tag{A.31}$$

where $\xi = \eta_i$ or $\xi = 1 - \eta_i$ depending on the considered node.

A.4 Pseudo-codes of the SSKF and SKF Methods

Algorithm 1: Pseudo-code of the SSKF method.

input : $\hat{x}_{k-1}$, u_k
output: $\hat{x}_k$

while *Algorithm is running* **do**
    if *Active area changes* **then**
        x^o_{k-1} is stored in the correct positions of x_{k-1}
        Redefinition of new x^o_{k-1} ;
    end
    Computation of $\eta_{1,k}$ and $\eta_{2,k}$ as in (5.16)
    Definition of $q_k(\eta_{1,k}, \eta_{2,k})$ as in (5.15)
    Computing K^{SS} as in (5.23)
    $x^o_k = x^o_{k-1} + K^{SS}_k (z_k - q^o_k x^o_{k-1})$
end
Updating of x_k considering x^o_k

Algorithm 2: Pseudo-code of the SKF method.

input : $\hat{x}_{k-1}$, u_k, P_{k-1}, R, $\mathbf{Q^o}$, Q^u
output: $\hat{x}_k$, P_k

while *Algorithm is running* **do**
    if *Active area changes* **then**
        x^o_{k-1} is stored in the correct positions of x_{k-1}
        $diag(\mathbf{P^o_{k-1}})$ and P^u_{k-1} are stored in a global vector P_{k-1}
        Redefinition of $\mathbf{P^o_{k-1}}$, P^u_{k-1}, x^o_{k-1}
    end
    Computation of $\eta_{1,k}$ and $\eta_{2,k}$ as in (5.16)
    Definition of $q^o_k(\eta_{1,k}, \eta_{2,k})$ as in (5.15)
    $\mathbf{P^o_{k|k-1}} = \mathbf{P^o_{k-1}} + \mathbf{Q^o}$
    $K^o_k = \mathbf{P^o_{k|k-1}} \mathbf{q^o_k}^{\mathbf{T}} \left(\mathbf{q^o_k} \mathbf{P^o_{k|k-1}} \mathbf{q^o_k}^{\mathbf{T}} + \mathbf{R} \right)^{-1}$
    $x^o_k = x^o_{k-1} + K^o_k (z_k - q^o_k x^o_{k-1})$
    $\mathbf{P^o_{k|k-1}} = (\mathbf{I_{4\times 4}} - K^o_k q^o_k)(\mathbf{P^o_k} + \mathbf{Q^o})$
    $P^u_k = P^u_{k-1} + Q^u$
end
Updating of x_k and P_k considering x^o_k, $diag(\mathbf{P^o_k})$ and P^u_k

A.5 The Dynamic System for the SKF Method

The system (5.36) is the local dynamic observable system. For this system, the split in (5.25) is still valid. The only modification is that the new observable part $\mathbf{P_k^{wo}}$ has n extra dimensions corresponding to the n:th order discrete filter. In the paper, a first order discrete filter is used, and then $\mathbf{P_k^{wo}}$ is 5×5.

$$\mathbf{P_k^w} = \left[\begin{array}{c|c} \mathbf{P_k^u} & \\ \hline & \mathbf{P_k^{wo}} \end{array}\right] \tag{A.32}$$

$$\mathbf{P_k^{wo}} = \left[\begin{array}{c|c} \mathbf{P}_k^o & P_k^{ndo\,T} \\ \hline P_k^{ndo} & P_k^{do} \end{array}\right] \tag{A.33}$$

where P_k^{do} is the scalar variance coupled to sensor dynamics and P_k^{ndo} is the vector 1×4 with the covariances between sensor estimation and observable parameters. $\mathbf{P_k^w}$ is a positive-semidefinite matrix, so is $\mathbf{P_k^{wo}}$. K_k is reordered to contain the unobservable part $K_k^u = 0$ and the observable part K_k^o, and the latter in the parameters related K_k^θ and dynamics related part K_k^s

$$K_k^w = \left[\begin{array}{c} K_k^u \\ \hline K_k^o \end{array}\right] = \left[\begin{array}{c} 0 \\ \hline K_k^\theta \\ \hline K_k^s \end{array}\right] \tag{A.34}$$

and (5.26) is now

$$\begin{aligned} \mathbf{P_{k+1}} &= \left[\begin{array}{c|c} \mathbf{I} & \mathbf{0} \\ \hline \mathbf{0} & \mathbf{I_{5x5}} - K_k^o H^w \end{array}\right] \left[\begin{array}{c|c} \mathbf{P_k^u} + \mathbf{Q^u} & \vdots \\ \hline \cdot\ \cdot\ \cdot & \mathbf{F_k^w P_k^{wo} (F_k^w)^T + Q^{wo}} \end{array}\right] \\ &= \left[\begin{array}{c|c} \mathbf{P_k^u} + \mathbf{Q^u} & \vdots \\ \hline \cdot\ \cdot\ \cdot & (\mathbf{I_{5\times 5}} - K_k^o H^w)(\mathbf{F_k^w P_k^{wo} (F_k^w)^T + Q^{wo}}) \end{array}\right] \end{aligned} \tag{A.35}$$

which shows that the SKF hypotheses are also applicable for the widespread dynamic system. $\mathbf{P_k}$ is a positive-semidefinite matrix so are $\mathbf{P_k^w}$ and $\mathbf{P_k^{wo}}$.

References

Ekstrand B (1983), Analytical steady state solution for a Kalman tracking filter. IEEE Trans Aerosp Electron Syst AES-19 6.

Hernández L (2004) Desarrollo de una Metodología para la Predicción y Optimización de Emisiones Contaminantes y Consumo en Motores Diesel de Automoción Mediante Redes Neuronales Artificiales. Ph.D. thesis, Universitat Politècnica de València. Departamento de Máquinas y Motores Térmicos

Hewer GA (1973) Analysis of a discrete matrix Riccati equation of linear control and Kalman filtering. J Math Anal Appl 42:226–236

Kalman RE, Bucy RS (1961) New results in linear filtering and prediction theory. J Basic Eng 95–108.

Lainiotis DG, Assimakis ND, Katsikas SK (1994) A new computationally effective algorithm for solving the discrete Riccati equation. J Math Anal Appl 3:868–895

Nash RA Jr (1967) The general solution to a second order optimal filtering problem. Proc IEEE 55:93–94

Schilling A (2008) Model-based detection and isolation of faults in the air and fuel paths of common-rail DI diesel engines equipped with a Lambda and a nitrogen oxides sensor. Ph.D. thesis, ETH-Zürich.

Simon D (2001) Kalman filtering. Embed Syst Prog 14(6):72–79

Sudano JJ (1995) Analytical solution for a steady-state Kalman filter tracker with random power spectral density process noise, vol 2. Proceedings of the IEEE National Aerospace and electronics conference, NAECON 1995:748–751

Turner JD (2009) Automotive sensors. Sensor Technology Series, Momentum

Curriculum Vitae

Dr.-Ing. David Blanco-Rodriguez
PhD Mechanical Engineer
dablarod@gmail.com
June 1984, Spain

Positions

2014–currently	**Project Manager**, *FEV Gmbh, Germany,* Research and Development in Passenger Car Diesel Engines.
2010–2014	**University Lecturer**, *Universitat Politècnica de València (UPV), Spain,* MsC Aerospace Engineering, advisor of 5 MSc Theses.
2009–2014	**Research Engineer** *CMT-Motores Térmicos Institute in UPV, Spain,* Engine Powertrains.
2011	**Research Engineer**, *University of Linköping, Sweden,* Vehicular Systems Department, Emissions Control and Modelling.
2007–2009	**Project Engineer**, *SENER, Spain,* Space Mechanisms and Structures Analysis.

D. Blanco-Rodriguez, *Modelling and Observation of Exhaust Gas Concentrations for Diesel Engine Control*, Springer Theses, DOI: 10.1007/978-3-319-06737-7,

Education

2009–2013	**International Ph.D. Mechanical Engineering**, *UPV, Spain*, Summa Cum Laude, *Dissertation: Modelling and Observation of Exhaust Gas Concentrations for Diesel Engine Control.*
2009–2011	**MSc Internal Combustion Engines**, *UPV, Spain* Avg mark: 9.3.
2006–2007	**MSc Aerospace Engineering**, *Universidad del País Vasco-Euskal Herriko Unibertsitatea (UPV-EHU), Spain*, Avg mark: 10.
2002–2007	**MSc Mechanical Engineering**, *UPV-EHU, Spain*, Avg mark: 8.7.

Publications

Totalling 6 journal articles (JCR), 7 conference papers, 1 educational paper. Highlights:

A Computationally Efficient Kalman Filter Based Estimator for Updating Look-up Tables Applied to NOx Estimation in Diesel Engines. Control Engineering Practice, 21(11) pp 1455–1468, 2013.

An On-engine Method for Dynamic Characterisation of NOx Concentration Sensors. Experimental Thermal and Fluid Science, 35(3) pp 470–476, 2011.

Distinctions and Memberships

2013	**UPV**, Nomination to 2015 Extraordinary Thesis Award.
2013	**Positive evaluation from ANECA** (Spanish Ministry of Education and Science) as a PhD holder Professor for public and private universities.
2012	**International Federation of Automatic Control (IFAC)**, Junior member of the TC Automotive Control.
2010	**Spanish ministry**, FPU grant for PhD candidates.
2002	**TECNUN University, Spain**, Awarded 1st in Mathematics National Contest MATENET.
2002	**Spanish Society of Physics**, Bronze medal in the National Challenge.
2002	**Sp. Soc. of Maths**, Awarded 1st in Mathematics Olympics in Basque Country.
2002	**Urdaneta High School in Loiu, Spain**, Graduated with honours.

Others

Languages	Spanish, English, German, Basque, Catalan.
Interests	Trail Running, Nordic, Mountain Biking, Music, Nature, Travelling, Reading.
Certifications	Open Water Diver, Official Children's Instructor.

MIX
Papier aus verantwortungsvollen Quellen
Paper from responsible sources
FSC® C105338

If you have any concerns about our products,
you can contact us on
ProductSafety@springernature.com

In case Publisher is established outside the EU,
the EU authorized representative is:
Springer Nature Customer Service Center GmbH
Europaplatz 3, 69115 Heidelberg, Germany

Printed by Libri Plureos GmbH
in Hamburg, Germany